VISUALIZING CULTURE

Visual
COMMUNICATION

Susan B. Barnes
General Editor

Vol. 4

The Visual Communication series is part
of the Peter Lang Media and Communication list.
Every volume is peer reviewed and meets
the highest quality standards for content and production.

PETER LANG
New York • Bern • Frankfurt • Berlin
Brussels • Vienna • Oxford • Warsaw

ROXANNE M. O'CONNELL

VISUALIZING CULTURE

ANALYZING THE CULTURAL AESTHETICS OF THE WEB

PETER LANG
New York • Bern • Frankfurt • Berlin
Brussels • Vienna • Oxford • Warsaw

Library of Congress Cataloging-in-Publication Data

O'Connell, Roxanne M.
Visualizing culture: analyzing the cultural aesthetics of the Web / Roxanne M. O'Connell.
pages cm. — (Visual communication; v. 4)
Includes bibliographical references and index.
1. Visual communication—Digital techniques. 2. Aesthetics. 3. Web sites—Design—Cross-cultural studies. 4. Web sites—Social aspects. 5. Mass media—Aesthetics. 6. Digital media—Social aspects. 7. Communication and culture. I. Title.
P93.5.O29 006.7—dc23 2014000508
ISBN 978-1-4331-2223-1 (hardcover)
ISBN 978-1-4331-2222-4 (paperback)
ISBN 978-1-4539-1290-4 (e-book)
ISSN 2153-277X

Bibliographic information published by **Die Deutsche Nationalbibliothek**.
Die Deutsche Nationalbibliothek lists this publication in the "Deutsche Nationalbibliografie"; detailed bibliographic data are available on the Internet at http://dnb.d-nb.de/.

Cover art: Paul S. O'Connell, all rights reserved. pauloconnell.com

The paper in this book meets the guidelines for permanence and durability of the Committee on Production Guidelines for Book Longevity of the Council of Library Resources.

29 Broadway, 18th floor, New York, NY 10006
www.peterlang.com

Printed in the United States of America

This is for Robbie

and for all my COMM 380 students

Contents

Foreword

The word "selfie"[1] was chosen to be the word of the year in 2013 by The Oxford English Dictionary. The word choice by the venerable OED draws our attention to how highly visual and mediated our world has become. Self-portraits have always been around. What is different is the ease and speed with which our lives can be visualized and communicated across time and space. The minutia of our everyday lives is now the stuff of an endless stream of tweets, posts, pins, grams, and vines. This over-pouring (and sometimes overpowering) digital documentation of our inner and outer cultural worlds into the vortex of social media is subject to culturally situated aesthetics codes, and rules. In other words, visual snapshots such as the "selfie" are not just individually generated messages but are inevitably culturally coded. This visualization of our culture raises several questions. How does culture mediate visual, digital design and practices? What does it mean to talk about a globalized, visual culture? How are our social subjectivities and cultural politics connected to our visual production and consumption? How do we interpret web visualities and imagery through cultural, cross-cultural and intercultural lenses?

Roxanne O'Connell's book, *Visualizing Culture: Analyzing the Cultural Aesthetics of the Web*, is a bold and concerted attempt to answer these questions.

It helps us understand the various ways in which technology is being used as visual platforms for information and design. It prompts us to think carefully about cultural universality and cultural relativity in relation to web design and visual media. It encourages us to appreciate the complex imbrications between visual aesthetics, cultural values, and global optics. But perhaps most importantly, *Visualizing Culture: Analyzing the Cultural Aesthetics of the Web*, alerts us to the fact that there has been a paradigm shift in the way we represent ourselves, engage the other, and communicate with the world. As well-known scholar Stuart Hall remarked, "history is constantly breaking in unpredictable ways," yet we believe that we "somehow go on being the same" (1997, p. 43). We have long privileged logocentricity and the primacy of the linguistic text. Consequently, much of how we understand and analyze visual imagery and visual web-based media stems from a linear, linguistic approach. In contrast, O'Connell's work urges us to consider how visual aesthetics, technology, culture and globalization are profoundly interlinked.

Both scholarly and trade writing on visual communication have predictably proliferated in the last few years. The scholarly work is typically heavily tilted towards the theoretical whereas the trade writing too often offers short-hand, cookie-cutter tips on visual communication. *Visualizing Culture: Analyzing the Cultural Aesthetics of the Web,* stands as a unique and much-needed addition to the growing literature on visual communication. O'Connell uses her extensive experience and expertise that has spanned the roles of manager/entrepreneur, usability engineer, researcher/author and educator to produce a book that is both practical *and* grounded in communication theory. The impetus behind this book is to provide well-developed and conceptually supported tools to examine and decode how visual aesthetics are mediated by cultural contexts. The idea of "seeing" as deeply cultural yet morphing through the vectors of a more universalized global economy is at the heart of *Visualizing Cultures.* O'Connell provides very specific and usable ways in which educators, scholars, and professionals of visual communication can traverse the cultural fields of this web visuality. Whether trying to develop pedagogical modules, research instruments, or practical solutions, this book offers systematic, lucid, and informed resources.

As a scholar who studies global media and identity, I am particularly intrigued by the way O'Connell knits the utilitarian with the theoretical. She states at the outset that *Visualizing Culture* is not about theory. I beg to differ. While this book may not wade through the thick and sometime impenetrable marsh of cultural and critical theory, it does frequently foray into some very meaningful conceptual conversations related to globalization and cultural identity. However, true to its purpose, the book does not linger long on untangling theoretical conundrums and swiftly moves on to elaborating its goal of helping us understand how cultures might potentially express themselves visually through the media. By giving the more practical focus of the book more heft *Visualizing Culture* promises to be an important touchstone work that engages a wide audience. As a reader, I know that I am now better armed to unpack the various cultural ramifications, iterations, and implications of the "selfie."

Anjali Ram
Barrington, Rhode Island

Notes

1 A self-portrait taken with one's smartphone or digital hand-held camera and usually posted on a social media site.

Acknowledgments

Everything comes from somewhere. When one begins to teach something, one goes first to what one has been taught. In addition to the writers and researchers mentioned in this work, I wish to thank my former teacher, colleague and friend, Bill Gribbons, whose globalization course got me thinking along these lines. I also want to thank the students of COMM 380: *Visual Media in a Cultural Context.* Their struggle to understand cultural filters and aesthetics was what pulled me to investigate better ways to explore and explain what we were seeing. They were, and continue to be, my research assistants, my interlocutors, and my motivation for writing this book.

I also wish to thank my long-time media ecology mentor, Susan Barnes, for inviting me to write this book. She does what a good editor must do if the work is to be valued: she probed and questioned, drove for clarity, argued for the reader, and, all the while, did this with efficiency and grace, helping me rethink the trouble spots and refine what might have been fuzzy.

I would like to express my appreciation to my colleagues on the Foundation for the Promotion of Scholarship and Teaching at Roger Williams University for providing the support needed for writing and publishing. A very special thanks goes to Kamille Gentles-Peart and Anjali Ram in our Global Communication program. They consistently provide the kind of nurturing environment that makes this kind of scholarship possible. Steeped in communication theory and practice, they are my guides and my dear, dear friends.

Finally, I want to thank my life-partner, soulmate and copy editor extraordinaire, Robbie O'Connell. He is my first reader and listener and the one who keeps me from losing my way within my own prose. If he can't follow what I'm saying, then I need to rewrite it and express it more clearly, cleanly, and succinctly. He keeps me on the path.

Roxanne O'Connell
Bristol, Rhode Island, 2014

PART I: Thinking about Patterns

Visualizing Culture is a *way finding* work. It is intended to help us recognize *where we are* on the Web based on visual cues that indicate the cultural aesthetic of that website and its author(s). We derive two benefits from knowing 'where we are.' First, we will see and listen differently when we are more finely tuned to the material before us. Second, in crafting our own messages for people situated in a particular cultural context, we will be better able to 'speak' to their 'listening.' As such, it falls into the category of a 'how to' book. It is not intended to impart deep theoretical and philosophical knowledge, although some of that is in here, but rather to provide practical methods for researching and analyzing what we see—and what might be at work underlying what we see—when we look at a webpage.

In short, *Visualizing Culture* is a book about Communication: communicating visually, whether it is within a particular cultural sphere or between two cultures or within an intercultural milieu where cultures intermingle. It is at this point that I must make an important and careful distinction. This is not a book about 'visual culture,' a field of study that combines elements of cultural studies, art history, anthropology, photography and the various philosophies and theories shared by these fields. Nicholas Mirzoeff, editor of the *Visual Culture Reader*, writes, "Visual culture is concerned with visual events in which information, meaning or pleasure is sought by the consumer in an interface with visual technology…." At first blush, it may appear that we are indeed working with visual culture here, especially when Mirzoeff's catalog of visual technologies include, "...any form of apparatus designed either to be looked at or to enhance natural vision, from oil painting to television and the Internet…." However, it is the last part of his definition that helps us understand why what you have in your hands is not a text on visual culture, per se. Visual culture takes a critical perspective on the visual image. "Such criticism takes account of the importance of image making, the formal components of a given image, and the crucial completion of that work by its cultural reception" (1998, p. 3). *Visualizing Culture* does indeed use some of the theories shared in the field of visual culture.

However, its intent is not critical but rather observational and analytical exploration. We use our eyes for exploring the many elements presented to us on the Web and we use what we know or can learn about a particular culture to think about and predict how what we see resonates with the people of that culture. We are *looking* at a culture, looking at the full range of visual cues used by that culture to represent its beliefs, values and attitudes, in such a way that someone from that culture will recognize that this webpage is for them.

Nor is this work about computer-mediated communication from the perspective of the technology. Indeed, the term 'computer-mediated' begins to date itself in this age when nearly everything with a screen, on our desktops and in our hands, is able to mediate culturally situated information. This book is about identifying elements within their cultural contexts that do a common thing. The focus in this work is on *functional* elements rather than the *ideological* message. It is about communicating the aesthetic that is culturally anchored—visual material that encompasses sensory and arbitrary information, spatially arranged and mediated through the Web, derived from the lived experience, values, beliefs, and behaviors of a particular community of people.

The real work in developing these analysis methods and the pattern language that emerge from them, as outlined in this book, began in 2003 when I first started teaching Visual Communication at my university. One of my assigned courses was entitled "Cultural Differences in Visual Media," later renamed to "Visual Media in a Cultural Context." With no further guidance than the title of a course which had never before been offered in our Global Communication program, I struggled with what studying visual media in a cultural context would mean. It clearly suggested 'seeing things' differently—and then deriving some meaning from that different view.

My first inclination was to look at film, but I soon realized that there was a richer pool of visual media coming from the World Wide Web. The challenge became how to find websites created by and for people of other cultures and how to interpret and understand what we, as seekers of these visual media, were seeing. The secret lay in my training as a usability engineer.

Usability as an art and craft has been around since the first tool was invented. Tools that are not usable waste energy and can cause harm—imagine a badly balanced hammer or a fire exit door that opens in and not out. With the dawn of the computer age in the mid-twentieth century, we shifted from direct interfaces, e.g., handles, buttons, grips, reins, to mediated interfaces, e.g., images, typeface, colors and shapes arranged to represent button-like links. With this shift from the concrete to the virtual, or representative, interface came an awareness that our perception is quite malleable to these new visual media. We had little trouble transferring our understanding of how these representations should behave. This led to an examination of how things in our environment, concrete or virtual, *tell us* about how they can be used—their *affordance.* Affordance, a term coined by J. J. Gibson in the 1950s, "refers to whatever it is about the environment that contributes to the kind of interaction that occurs" (Greeno, 1994, p. 338). Usability became a science.

My basic instincts for trying to understand what we are seeing on the Web were initially developed with computer hardware and software many years ago when I began working with the team at Digital Equipment Corporation's Software Usability Engineering group. There I met Dave Marca, John Whiteside, Chauncey Wilson and Dennis Wixon, pioneers in human-computer interaction and usability. I was, at the time, a systems analyst and organizational consultant interested in how people dealt with working in technical environments. For several years, I had the opportunity to work with, and learn from, some of the finest minds in the technology/usability world. In 1992, when Digital began its radical downsizing efforts and tens of thousands of us were being laid off, the Web was just being born and some of us, fortunately, found ourselves working in this new and exciting area of technology.

In the following years, as a web designer, usability engineer and, later, as a teacher of visual communication, the many tools and skills I had accumulated since those days at DEC translated to ideas, 'patterns,' I could bring into my practice and into the classroom. These ideas, and those that my students and I discovered together over the past decade, have evolved into this notion of cultural aesthetic patterns and the language we can use to communicate them.

A word about the audience for this book is perhaps in order. There are four possible audiences for this text. Educators may find it a useful methods text. As visual communication and intercultural communication programs at universities proliferate in our visual media saturated landscape, this text can be used in a research methods course that overlaps these areas of study. Likewise researchers and scholars of digital and global media can use this pattern language to develop codes that track reception and cultural influence. 'Digital' includes traditional HCI as well as the emerging fields of digital media that are not computer based such as mobile devices and media appliances.

In the commercial professional realm, nothing is quite as challenging in our new global communication landscape than creating messages for cultures not our own. The creation of visual messages is particularly difficult because so much of what a person takes in is not concrete or consciously acknowledged. Digital communicators and message creators can use the approach outlined in this book to research and better understand what visually resonates with certain audiences at a cultural identity level. Both Web and graphic designers may find the tools and the pattern language outlined here as a useful references when undertaking projects that involve cross-cultural, globalized or localized design challenges.

Nothing beats experiencing the patterns we are developing in some systematic way. To that end, I have included, in Part 2, exercises entitled "Putting Patterns into Practice." Some of these exercises, like Affordance Testing are part of a usability engineer's toolbox and generate qualitative and quantitative data that become part of a culture's aesthetic 'big picture.' Others are means by which the researcher/designer can experience cultural values, attitudes and behaviors that are not his or her own. These also generate data, but the true value is in achieving the widened perspective of having been in another's milieu for a bit.

As I explained at the beginning of this section, this is a practical 'how to' book. However, if, in using this book, you do have deep theoretical or philosophical insights into how cultures respond to visual cues on the Web, and are then able to use that information to make communication within and among cultures better, we will both, you and I, be profoundly pleased and satisfied.

Chapter 1

Introduction

> *As interworked systems of construable signs..., culture is not a power, something to which social events, behaviors, institutions, or processes can be casually attributed; it is a context, something within which they can be intelligibly—that is thickly—described.*
> —Clifford Geertz, *The Interpretations of Cultures*

Visual media are windows onto societies and their culture. This has long been true of print imagery, television, film and the Web. From the beginning of the Web in the 1990s, designers have employed patterns to define and create a website's structure and utility—menu ribbons, login forms, blog layouts, wikis, news formats (headline, snippet, link for more.) But there other patterns—subtle visual cues that 'speak' to the cultures to which these websites are directed and within which they have been built. This chapter sets forth the context for this book in terms of the concept of a 'visual cultural aesthetic,' how one identifies this aesthetic and how it can be used. This chapter also outlines the methods by which we may develop a 'pattern language' that help us identify aesthetics in websites that are culturally derived and to which we expect audiences from those cultures to respond.

The idea of a 'pattern language' is not new. In 1999, Jennifer Tidwell published an online guide for a pattern language for human-computer interface (HCI) design (2011). Ian Graham (2003) wrote *A Pattern Language for Web Usability*, or *wu*. All three, Tidwell, Graham, and this book, acknowledge the work of Christopher Alexander, Sara Ishikawa and Murray Silverstein in their seminal book, *A Pattern Language: Towns, Buildings, Construction* (1977), and for the overall concept of creating a lexicon that helps us identify archetypal elements that people of a particular culture recognize. Their pattern language addressed the issue of building environments in which people live, work, learn and play. They, too, drew from cultures all over the globe to help them define concepts such as 'density rings' (pattern #29), 'the sitting circle' (#185), and the 'roof garden' (#118), as well as more personal patterns such as 'waist-high shelf' (pattern #201), 'built-in seats (#202) and my personal favorite, 'child caves' (#203).

Visualizing Culture extends and echoes that concept by creating and identifying terms we can use to describe the cultural aspects of Web design and digital communication. Language is 'plastic' in that it continually evolves to meet the needs of those using it. While words and terms often retain their original meaning, new meanings become attached to words to help us describe new contexts and phenomena, particularly in the realm of digital technology. For example, as someone who has studied Heidegger and metaphysics in college, I was surprised to see the term 'ontology' show up when I was working for Digital Equipment Corporation and again when I did graduate work in the area of information science and usability. What was first (and still) used to describe categories of being and existence came to describe knowledge domains as they are categorized in terms of properties and relationships. Just as the term ontology was borrowed from philosophy to be used in computer information science, we may find ourselves borrowing terms and ideas from philosophy and other fields to help us describe phenomena in both the cultural and technological domains that answer questions such as: What constitutes the identity of this culture? What are the characteristics of the features in this website? How are they essential expressions or representations of this culture? Or are they accidental appropriations?

The areas from which we borrow the concepts and words to build our pattern language include, but are not limited to, design, technology, semiotics, cognitive science, and anthropology. In many cases, the essential or root meaning of the concept or word will not be changed, but the context in which it is applied will center on what we are seeing in the digital media landscape. Just as the map is not the territory, our concepts and words are metaphorical in nature and should not be used to essentialize any culture.

It is also useful here to invoke Stuart Hall's concept of encoding and decoding. By using the term 'code' he does not mean 'computer code' or HTML but rather the signs or sign-vehicles that are produced for user consumption.

Current Literature

Exploring a culture visually is not the purview of any one discipline. It is an interdisciplinary endeavor. As Roland Barthes advises, interdisciplinary work:

> ... [does] not merely confront already constituted disciplines (none of which, as a matter of fact, consents to leave off). In order to do interdisciplinary work, it is not enough to take a "subject" (a theme) and to arrange two or three sciences around it. Interdisciplinary study consists in creating a new object, which belongs to no one. (1972, p. 72)

Therefore the works that contribute to our understanding of the cultural aesthetic encompass many disciplines and areas of study.

There are several texts that cover research methods for visual media. Most of the literature treats with traditional (non-digital) media. The current foremost visual research methodology text is *Visual Methodologies* by Gillian Rose (2001, 2007). However, Rose, writing at the turn of the twenty-first century, quite specifically, and rightly, chose to not engage the digital sphere. At that time, the Web itself was not even a decade into its existence and no one could anticipate the changes that would be brought about by Web 1.0 and Web 2.0 technologies, and the even more dramatic changes in the visual 'mediascape' introduced by smartphones and iPads. In 2001, if you wanted to see digital material digitally, you had to have a computer—and a great deal of patience as download speeds were tortuous at best.

The methods covered in Rose's text are useful and valid for traditional visual criticism research. Two methods in particular, content analysis and audience studies, also prove useful in digital media if modified for the particular mediascape of the Web. Audience studies become 'user' studies when we move to the Web. Television, as Marshall McLuhan would have to acknowledge, is a passive experience compared to the Web (2002, pp. 22–32). The Web is largely interactive. Television is composed of moving images. The vast majority of the Web consists of still images or pages that users 'move through.' Content, and its composition, is also something that can be measured. Most websites consist of both words and pictures. We can look at words and images as two different modes of content. Layered on content composition is the purpose of those words and images: are they navigational, decorative, informative? We will discuss various content analysis techniques in Chapter 5 when we look at content modalities.

To map, explore, analyze—and use—knowledge about the digital visual mediascape in a cultural context, we need to take a *bricolage* approach—in other words, use what we have available from several disciplinary areas and analytical methods to fashion, as the constructivist Seymour Papert would put it, a problem solving approach by trying, testing, and playing around: blending methods we have and bending them to our purpose.

We start by identifying what it is we want to know: what are the visual cues with which a particular cultural audience identifies? For example, how do southern Italians know that the website they are visiting is targeted at them? What imbues it with *Italianicity*, as Barthes (1977) would put it?

We then use those traditional visual research methods mentioned above and combine them with methods from other disciplines—anthropology, cognitive psychology, design—hybridizing some to better meet our needs, forming, in the process, a pattern language that helps us to build a cultural aesthetic profile.

So why isn't this book a catalog of the profiles created while researching the cultural aesthetic over the past eight to ten years? Because the digital visual technologies and their platforms are plastic, not static. Whatever we discover today may not entirely hold true a decade hence. However, if we learn how to 'see' the cultural aesthetic cues and patterns and use those methodological tools, we can identify when there are shifts in the aesthetic or in emergent audiences that respond to a different aesthetic. As John Berger writes in *Ways of Seeing* (1972), "It is seeing which establishes our place in the surrounding world; we explain the world with words, but words can never undo the fact that we are surrounded by it. The relation between what we see and what we know is never settled" (p. 1–cover). If we are scholars, we can understand and explain emergent cultural influences. If we are message creators, we can insure that we are visually crafting our messages to better resonate with their intended audience.

In addition to Rose's *Visual Methodologies*, other texts that look at analyzing visual media or which propose theories of construction or reception include *Designing Across Cultures* (Lipton, 2002), *Reading Images: The Grammar of Visual Design* (Kress & van Leeuwen, 1996), *Multimodality* (Kress, 2010), and *Visualizing the Web* (Josephson, Barnes & Lipton, eds., 2010). These works tend to speak to areas and aspects of visual analysis, however, they do not attempt to establish a pattern language by which we can communicate a shared understanding of intersubjective criteria and observations used to analyze the cultural aesthetic of digital media. Singh and Pereira's *The Culturally Customized Website* (2005) comes close to defining cultural aspects of Web media. It is intended to help designers create culturally customized websites. However, it relies heavily—almost exclusively—on the work of Geert Hofstede, a researcher hired by IBM in the '60s and '70s. A largely formulaic and prescriptive text, it seeks to make recommendations for designing new websites based on many examples from shortly before its publication, but is not an approach to analytically explore what has come into being since. It does include many of the charts and tables first published in Hofstede's *Cultures and Organizations*, contextualized for the five value pairs we also explore here.

While this work, *Visualizing Culture,* also calls on Hofstede's work, it tempers his theories with more current understandings in terms of diversities not reflected in Hofstede's original work. A good example of this is how Hofstede used the terms 'masculinity' and 'femininity' to define a cultural value dichotomy. Using a gender term to describe these values occludes the cultural attitude we are seeking to identify because Western 'gender' biases dominate. I prefer to use terms that more closely describe what we see exhibited in the visual media we are examining. However, I also break these down into subtler distinctions informed by similar Jungian value pairs: rational/emotional, objective/relational, materialistic/quality of life. In other words, the richer our pattern language, the better we are able to characterize a culture's aesthetic.

Insofar as any study of visual patterns requires some visual literacy, particularly regarding perception and cognition, we must include a brief examination of Gestalt phenomena and basic semiotics. There are several Gestalt principles, explanations of visual perceptual phenomena that help us predict how we will interpret what we see based on color, shape, size, texture, orientation and spatial arrangement. Common phenomena include the principles of proximity, continuance or continuity, similarity, common fate, closure and the figure/ground articulation. A good designer will be familiar with these principles and use them to cognitively exploit our tendency to group things that are close together, or that are similar in shape and size, for instance, headlines and teaser copy or the buttons in a menu ribbon. Icons, those over-simplified graphic representations of objects or functions that are commonly used in computer and web applications and sites, exploit the Gestalt principle of closure, where we 'see' what isn't actually there because we have enough information from what is suggested to complete the image in our mind's eye (Todorović, 2007).

While Gestalt phenomena are largely biologically derived and universal in scope, semiotics is deeply connected to culture and representation. According to Swiss linguist and semiotician, Ferdinand de Saussure at the beginning of the twentieth century, semiotics is the study of signs, symbols and their signification as elements of communication. He described it as "a science which studies the role of signs as part of social life" (1983, pp. 15–16). Gunther Kress and Theo van Leeuwen, in *Reading Images: The Grammar of Visual Images* (1996), define their work as 'social semiotics' emphasizing that, "Communication requires that participants make their messages maximally understandable in a particular context," and that they "choose forms for the expression of what they have in mind, forms which they see as most apt and plausible in the given context" (p. 11). As an approach for our pattern language, social semiotics is very useful. "It is a form of enquiry. It does not offer ready-made answers. It offers ideas for formulating questions and ways of searching for answers" (Van Leeuwen, 2005, p. 1).

Visualizing Culture teaches us how to see and analyze visual digital media regardless of how they evolve, or, as the saying goes, it teaches us how to fish. It calls on multiple anthropological, visual, cognitive and design theories and, in the process, establishes a process by which we can develop a way to communicate what we know about cultures and aesthetic preferences in order to better understand what we see and, thereby, better formulate effective visual messages.

One might ask, "Why a pattern language? Why not a book of rules or heuristics?" A fair question—heuristics work well when we are talking about basic usability, that which is biologically or physiologically situated and relatively unchanging. But there is no one set of heuristics for culture. Culture is complex. It is, at one and the same time, traditionally rooted and ever evolving: anchored in place and carried to new communities as part of one's ethnoscape. Our own culture defines us and is defined by us. The member of a given culture knows what fits and what doesn't.

Because, in this work, we are looking at culture through a largely semiotic lens, looking for signs and cues and probing for their meaning, I look to Clifford Geertz for a definition of 'culture.' Ironically, long before the personal computer and the World Wide Web were invented, Geertz used the same metaphor to talk about culture:

> Believing, with Max Weber, that man is an animal suspended in webs of significance he himself has spun, I take culture to be those webs, and the analysis of it to be therefore not an experimental science in search of law but an interpretive one in search of meaning. It is explication I am after, construing social expressions on their surface enigmatical. (1973, p. 5)

Geertz's explication takes the form of 'thick description,' situating what is observed into a particular context, and then describing it, giving, not just the denotative facts of the observation, but also all of the connotative meaning that can be discovered by exploring the context in which the observation and its meaning exist. "The whole point of a semiotic approach to culture is…to aid us in gaining access to the conceptual world in which our subjects live so that we can, in some extended sense of the term, converse with them" (p. 24). It is in thick description that we will find useful patterns—patterns that don't necessarily lead to rules and guidelines, but rather to a deeper understanding of culture as context.

As I have said, there are many pattern languages. The purpose of this particular pattern language is to help us *see* how a culture represents itself in all its many flavors. And in coming to understand these flavors and the contexts in which they appear, we can better converse with the people in those cultures.

Using this Book

This is a book about the visual. Not surprisingly, there are many visual examples within, however many of them are of webpages that have a native resolution of 72 dots per inch (dpi). Resolution for paper is more than four times higher, so some of the images here may appear a bit blurry. If our intent was to show the reader the content of a webpage in all its detail, this would be a problem. Fortunately, the actual content of most of our examples is largely irrelevant at that level of detail. The words or the images on the page, except in how they help us distinguish between information versus advertising content, are not the focus. What is important is that the reader sees the impressions of color and ratios of visual to verbal material in the aggregate, and how using a montage approach, or superimposing a thirds grid, can help us see those better. In other words, where applicable, the image of the webpage is intentionally lower in resolution.

One of the challenges in writing about culture, technology, and design (aesthetics) and their function in visual communication, is the extent to which all of these intersect and overlap. To isolate patterns belonging to one sphere cannibalizes from the others making them appear less important or 'rich.' Better to think of all three of these: culture, technology and design, as threads of meaning that run through all of the chapters in this book, allowing the chapters themselves to provide focus on a particular aspect of the patterns in our pattern language. Chapter 2 lays the groundwork for just such a weaving. It begins by looking at the basic provenance and function of pattern language and goes on to establish how this work defines the terms 'culture,' 'technology,' and 'design.' At the end of each subcategory, a list of pattern terms maps their connection to one or more of the dominant root concepts, as determined by the primary research tools associated with those concepts. This provides the reader with a non-linear way to find the pattern terms and methods they are looking for. *Way finding* is itself an important pattern in the websphere. One might suggest that it is the *root* pattern, as it is the Web's ability to take us to all parts of the world that makes it what it is.

Getting started with an observation project is often fraught with chicken or egg decisions. Chapter 3 discusses two approaches, the first using the evidence provided by the websites, the second starting with what we know about the target user. It follows with data collection and recording methods and includes, at the end of the chapter, a cultural aesthetic grid, to be used with each website, where observations can be documented. The grid itself is based on a similar tool used in a Globalization/Localization course taught by Bill Gribbons at Bentley, and modified to include the pattern language criteria outlined and discussed in this book.

Chapter 4 examines various color palettes that appear to dominate the mediascape of various cultures. By noticing certain trends and patterns, communicators can identify not just the colors that resonate with an audience but also their intensity, temperature and symbolism. A great deal of research has been done regarding the symbolic use of color in different parts of the world. This book does not catalogue these but rather provides tools for establishing which colors in a Webpage carry symbolic significance and which are simply functional. This can then be cross-referenced with the existing research to validate your findings.

Chapter 5 looks at content—not the specific words or images which may change from day to day or hour to hour, but the content's modality. A website's 'modality' indicates which information mode is privileged for that culture: images or words. We also need to determine which function the images or words fulfill: decoration, navigation or information. In dealing with content, one must also look at its spatial orientation. Obvious directional attributes, such as reading direction (left to right, right to left, vertical) are ready clues to a culture's reading pattern. However, content density, or the amount of white space on a Webpage, is also an important cultural aesthetic. The content aspect of a website is not something that is easily measured. It interacts with several other measures such as static or dynamic, quantity and type of imagery, and represented values. The first step in examining those is an extensive visual content analysis. Therefore, this chapter outlines the use of an analytical grid and the information we can learn from it.

Deeply threaded into our cultures are patterns of time—mode and tempo: two related but distinct aspects of a website's basic design structure. Chapter 6 explores the relationship between culture and time, looking at both the philosophical and the psychological dimensions as they pertain to culture. For our pattern language we borrow from Edward T. Hall's observations as presented in *The Dance of Life* (1983) and *Beyond*

Culture (1976), to look at time modalities and a culture's orientation toward time. In *A Geography of Time* (1997), Robert Levine expanded on Hall's research studying the relationship of various cultures by adding 'tempo'—looking at how fast people walk, how long transactions take at banks, post offices, train stations, and other temporal idiosyncrasies of culture. These relationships to culture are not always readily apparent in websites, but combined with examinations of other visual media, patterns can be perceived revealing subtleties that might attract or repel the target audience.

Whether a culture prefers interactive or passive websites is captured in patterns of *audience interaction* and is discussed in Chapter 7. We see this most often in multinational sites where the U.S. created site has a great deal of Flash™ and lots of elements with which users can interact. When we compare localized versions of these sites and see a decrease in interactivity, we can speculate whether this is a technical consideration driven by bandwidth or an aesthetic one, indicating a cultural preference for stability or linearity. Either way, coding audience interaction elements provides important information for the researcher and designer.

Chapter 8 explores deeply cultural perspectives concerning a website's content. As with the previous chapter, we examine cultural cues concerning imagery (what is in the images), context (their meaning), and attitude (how they express it). These highly nuanced visual cues are those that deeply resonate with the intended audience. Much like using local idioms in a verbal language bridges the formality gap, understanding these visual cultural cues helps the audience 'know' that this is for them. It is in this chapter that we draw upon Hall and Hofstede's work and further refine it based on richer and less chauvinistic cultural distinctions. Hofstede's work, in particular, tended to focus on a cultural milieu that was, at the time, dominated by Western gender roles. While these gender-based presumptions about behavior and cultural expectations may still linger in some parts of our society, we no longer assume that they hold true for all cultures. This chapter draws on the work of E. T. Hall to look at high and low context cultures, and extends and melds the works of Hall (1959, 1976, 1983, 1987), Hofstede (1997) and Rokeach (1968, 1973) to define cultural value pairs such as rational/emotional, objective/relational, and materialistic/quality of life.

Chapter 9 centers on the technology of the Web itself and its cultural impact. A website's usability extends beyond mere physiological patterns, although these are important. Of greater concern for the purposes of this pattern language is how a culture interprets what the Web offers. The concept of uncertainty avoidance is subtly linked to a website's usability performance. This chapter outlines how to examine the usability of a website, in conjunction with the other factors already discussed in the previous chapters, to understand the target culture's tolerance for ambiguity. A culture that values 'saving face' will have a low tolerance while another culture, for instance the gaming culture, may have a very high tolerance for errors as they contribute to learning and winning the game.

No pattern language is successful without exemplars that help us see how these patterns can be combined and used to either deconstruct an existing thing or plan and create something new. Chapter 10 gives several examples where this method was used to identify basic cultural patterns and extrapolate a profile for producing more focused visual communication or for identifying emerging cultural trends. With close to 2000 examples to pull from, the material is rich. But is not the intention of this book or this chapter to codify or suggest that conclusions regarding certain cultural patterns are fixed. Rather, it is the work of this chapter to show readers how to use this pattern-based method to do their own investigating and exploring. A few key examples of both functional and cultural patterns are provided to add clarity to the whole idea.

The Afterword provides philosophical musings on globalization and its impact on cultural aesthetics and suggests some areas for research and further exploration and expansion of the pattern language itself. Will cultural aesthetics disappear as Web 2.0 technologies proliferate and templates homogenize the visual landscape of the Web? As new technologies and platforms emerge that are more visual, how will cultures that are verbally oriented respond? We are already seeing a shift in the privileging of images over words with the introduction of the iPad and other intensely visual devices. The test of a good pattern language will be how well it adapts to emerging technoscapes and evolving cultural aesthetics.

Summary

This chapter defined the scope of *Visualizing Culture* providing a rationale for developing a pattern language and outlining various applications wherein a pattern language would be useful. It also summarized relevant current literature that the reader may use to further explore specific background areas and disciplines that contribute important terms, concepts and vocabulary to our pattern language. Finally, this section provided an overview of the book and what material can be found in each chapter as a guide to the reader.

Chapter 2

A Pattern Language: Culture, Technology & Design

> *[T]he feeling a people has for life appears, of course, in a great many other places than in their art. It appears in their religion, their morality, their science, their commerce, their technology, their politics, their amusements, their law, even in the way they organize their everyday practical existence.... expressions of human purpose and the pattern of experience they collectively sustain.*
>
> —Clifford Geertz, "Art as a Cultural System"

What is a pattern language? Before we get to 'language' we must first talk about patterns. The word pattern comes from *patron,* the French word for template. We not only recognize patterns, we seek them out. It is part of our cognitive process. Pattern-seeking and pattern-recognition help us make sense of our world, especially our visual world.

When the Web was first developed in the early 1990s, the patterns most often used were taken from the only other visual information form we knew: 'words' and, when download bandwidth allowed, 'images,' arranged by 'pages.' 'Words,' 'images' and 'pages' are very basic elements in the Web's pattern language. Without words and images, a Webpage is merely a blank space. As evidence of the power of patterns, when we call up a Webpage and see only a blank space, we immediately register that something is not right—we do not interpret the lack of words or images as an extreme zen-like design of white space or as a specific functionality ("This page was intentionally left blank"—a paradox as the printing of those words negates the blankness of the page.) No, when we see nothing on the page, we assume that there is a problem because a blank Webpage does not fit the pattern we expected.

In the early days of the Web, programming and display constraints meant that everything was pretty much white or gray, often centered and 'wall-to-wall,' or from one side of the browser window to the other.

People learned to create Webpages by downloading the source code for an existing Webpage and deconstructing its HTML (HyperText Markup Language) script to understand how each tag rendered the desired pattern. Browsers like Netscape made it easy for users to view the source code of a page (Figure 2.1) and a little detecting and messing around with changing the code and then viewing it in a browser helped people learn how to make their own webpages. The Webpage as raw code, in this instance, was itself a pattern to be used over and over again, switching out the information content and reusing the format. Note the early adoption of blue underlined text to indicate Hypertext links (Figure 2.2).

Figure 2.1 - Raw HTML as seen by viewing the Source code

```
<TITLE>What's New, June 1993</TITLE>
<H1>What's New, June 1993</H1>

<DL>
<DT> June 27, 1993
<DD> Digital Equipment Corporation is running a
         <A
         HREF="ftp://gatekeeper.dec.com/hypertext/gatekeeper.home.html">Web
         server</A> from their <A
         HREF="ftp://gatekeeper.dec.com/">anonymous FTP server</A>.
         Included are pointers to <A
         HREF="file://gatekeeper.dec.com/archive/pub/DEC/DECinfo">DEC
         product information</A>,
         <A HREF="file://gatekeeper.dec.com/archive/pub/DEC/dec-faq">Ultrix
         and OSF/1 FAQ's</A>, <A
         HREF="ftp://gatekeeper.dec.com/hypertext/cra-reports.html">DEC
         research reports</A>, and more. <P>

         The Army Research Laboratory is now running a <A
         HREF="http://info.arl.army.mil/">Web server</A>.  Included is
         information on <A
         HREF="http://info.arl.army.mil/SciVis/scivis.html">the ARL
         Scientific Visualization effort</A>, a picture of <A
         HREF="http://info.arl.army.mil/Gif/KSR2.gif">the arrival of a
         KSR-1 supercomputer at ARL</A>, and more. <P>
```

Figure 2.2 - HTML from Figure 2.1 rendered in an early Netscape browser

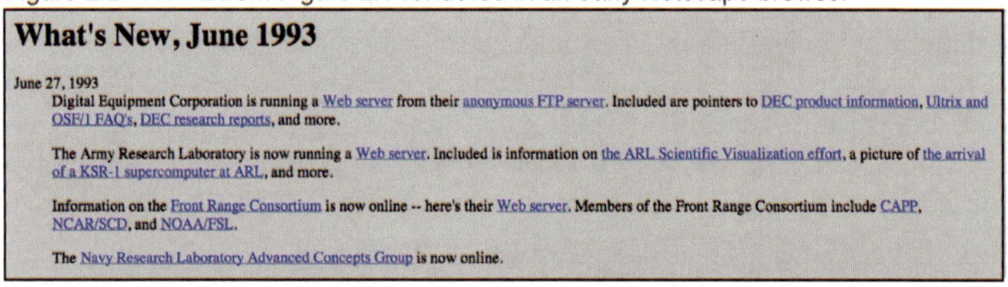

What's New, June 1993

June 27, 1993

Digital Equipment Corporation is running a Web server from their anonymous FTP server. Included are pointers to DEC product information, Ultrix and OSF/1 FAQ's, DEC research reports, and more.

The Army Research Laboratory is now running a Web server. Included is information on the ARL Scientific Visualization effort, a picture of the arrival of a KSR-1 supercomputer at ARL, and more.

Information on the Front Range Consortium is now online -- here's their Web server. Members of the Front Range Consortium include CAPP, NCAR/SCD, and NOAA/FSL.

The Navy Research Laboratory Advanced Concepts Group is now online.

As users demanded better aesthetic fidelity to those visual forms with which we were already familiar, new enhancements to the Web's markup language, its programming code, were developed giving us margins, frames, vertical and horizontal layouts, faster loading images and the ability to wrap words around them. New patterns emerged that indicated functionality: buttons and links became navigational elements anchored in menus displayed either as a ribbon across the top or as a table of contents going down the left or right side margin of the page. 'Menu,' 'button,' 'ribbon,' 'margin,' 'table of contents' all became new terms in the Web's pattern language. These evolutions were driven by the patterns and pattern terms found in media such as books, newspapers, magazines—three-dimensional information and entertainment objects from our physical world we wanted to see represented in the Web's virtual world.

This was possible because a pattern is not the object itself but rather a collection of perceptual attributes that different media use to communicate by exploiting our familiarity with the pattern. And patterns change and evolve. Buckminster Fuller writes:

> A pattern has an integrity independent of the medium by virtue of which you have received the information that it exists. Each of the chemical elements is a pattern integrity. Each individual is a pattern integrity. The pattern integrity of the human individual is evolutionary and not static. (Fuller, 1975, p. 228)

So, back to the question—what is a pattern language? For our purposes, two analogies come to mind. The first is from *A Pattern Language: Towns, Buildings, Construction* (1977). In this work, Alexander, Ishikawa and Silverstein develop a collection of patterns that describe a building environment problem and a core solution "in such a way that you can use this solution a million times over, without ever doing it the same way twice" (p. x). Patterns are inherently multi-use but they are also interdependent:

> In short, no pattern is an isolated entity. Each pattern can exist in the world, only to the extent that is supported by other patterns: the larger patterns in which it is embedded, the patterns of the same size that surround it, and the smaller patterns which are embedded in it. This is a fundamental view of the world. It says that when you build a thing you cannot merely build that thing in isolation, but must also repair the world around it, and within it, so that the larger world at that one place becomes more coherent, and more whole; and the thing which you make takes its place in the web of nature, as you make it. (Alexander et al, p. xiii)

The second analogy comes from another form of construction, but one with which we might all be familiar: clothing. A clothing pattern is made up of many elements, depending on the complexity of the garment. A pattern for a shawl would be quite simple as most shawls are long lengths of woven or knitted cloth. A pattern for a tailored jacket would be much more complex, composed of pieces for the back, the front, perhaps the sides, lapels, the collar, sleeves, pockets and lining. All jackets have a front, back and sleeves as do dresses, shirts, tops, etc. It is the function and the style that contribute to the greater pattern. But without the pattern *language* of front, back and sleeves, long and short, wide and narrow, we would spend a great deal of time communicating the function of the garment and each element in it. Function and style are largely culturally derived. A sari is a long strip of cloth—up to nine yards in length—that is intricately folded and fastened around the waist leaving a long end, the pallu, to drape over the head and/or shoulder. It is not simply a shawl. A Nehru jacket is a style considerably simpler in construction than a formal tuxedo jacket or an anorak. It does not have the lapels of a tuxedo jacket or the fur-lined hood of the anorak. Each of these jacket styles uses *some* but not *all* of the patterns that work for a jacket. A tuxedo jacket would never have a hood, fur-lined or otherwise.

The power of a pattern language is that it helps us identify those culturally influenced differences between a shawl and a sari, between a tuxedo and an anorak. These differences are a combination of culture and design and, in some cases, technology. Each of these areas are explored in more detail below, introducing us to ideas and theorists whose work we will draw on for our pattern language. At the end of each exploration is a list of the specific concepts, the theorist upon whose work the concept is built and the subsequent chapters in which the patterns can be found and more fully explained and illustrated.

Culture

How does a culture represent itself on the Web? Do the visual components of a website, be they navigational, informational or decorative, accurately represent a culture? Does the global nature of the Web contribute to a homogenization of cultural aesthetics? Or do designers and audiences assert a cultural aesthetic after new homogenizing technologies become established? These are questions that require 'thick descriptions' (Geertz, 1973, pp. 3–30) and a basic understanding of what we call 'culture,' 'representation,' 'aesthetics' and 'technology.' We also need to reflect on the changing, evolving and interdependent nature of each of these concepts.

Stuart Hall (1997) tells us that, "culture is concerned with the production and exchange of meanings—the 'giving and taking of meaning'—between the members of a society or group" (p. 2). Edward Hall writes, in *The Silent Language* (1959), that culture is "the way of life of a people, the sum of their learned behavior patterns, attitudes and material things" (p. 31). Geert Hofstede, in *Cultures and Organizations* (1997) uses the computer as a metaphor to describe how patterns of thinking, feeling and acting turn into 'culture'. He declares, "It [culture] is the collective programming of the mind which distinguishes the members of one group or category of people from another. Culture is learned, not inherited. It derives from one's social environment, not from one's genes" (p. 5). Clearly, cultures are made of patterns: social, behavioral, attitudinal. If cultures are composed of patterns, then we can develop a pattern language that helps us identify what visually speaks to a particular culture, what resonates with the members of that culture.

More enigmatic, and yet profoundly insightful, is Marshall McLuhan's frequent observation about culture, media and technology: whoever discovered water, it wasn't a fish (1968, p. 175). Our own culture is all around us. It is the metaphorical water in which we swim, the very air that we breathe. It isn't until we are removed from our culture, a 'fish out of water' so to speak, that we realize what it is and how it might differ from other cultures. And it is in our modes of communication that we represent ourselves and our culture. James Carey, a pioneer in bringing a cultural perspective to the study of communication, proposes a culturally ritualized view of communication in *Communication as Culture* (1989). He suggests that communication is "directed not toward the extension of messages in space but toward the maintenance of society in time; not the act of imparting information but the representation of shared beliefs" (p. 18). This perhaps comes closest to the overall approach our pattern language wishes to reflect. Above all, a Website targeted at a specific audience seeks to create a visual atmosphere that resonates with the aesthetic expectations of that audience.

Critics of the work of some of these sources, in particular the work of E. T. Hall or Hofstede, may make the claim that their ideas about culture are outdated. That can certainly be debated. However, in creating our pattern language, we are seeking words and terms that describe what we are seeing. To do that, we may have to re-describe the meanings of some terms as they relate to the development of this pattern language. As Elizabeth Wuertz (2005) found in her research, "… the qualitative study of local[ized] websites around the world… shows that in spite of these important criticisms, communication patterns today still resonate with the cultural dimensions proposed decades ago" (n.p.). Just as Alexander's building patterns are interdependent, cultural patterns are complex, organic, integrated within themselves, and interdependent with any given culture's history, environment, and relationships. What we see in any cultural self-representation is just the tip of that culture's iceberg. However, it does give us clues to what lies beneath the water. While the patterns we identify are not the cultures themselves they are visual evidence, of the culture's self-representation. They are the symbolic process embedded in how cultures visualize themselves and then represent themselves to others within their cultures and without.

Cultural anthropology patterns

- Attitude: rational/emotional (Hall/Hofstede), Chapter 8
- Context: High/low (Hall), Chapter 5
- Identity: collectivist/individualist (Hofstede), Chapter 8
- Identity: objective/relational (Jung/Hofstede), Chapter 8
- Information flow (Hofstede), Chapter 5
- Time mode (Hall), Chapter 6
- Time tempo (Levine), Chapter 6
- Value orientation: materialistic/quality of life (Hall/Hofstede), Chapter 8

Technology

Media ecologists[1] maintain that the invention of language was our first communication technology (Donald, 1991; Logan, 2007; McLuhan, 1964). What is often overlooked is that language from the outset was always multimodal: the mimetic or gestural, the oral or spoken, and the visual. Our first visual information forms were maps, perhaps drawn in the dirt depicting the path from one village to the next or to the place where food could be found. The caves of Chauvet, Lascaux and Altamira are purported to have astronomical maps embedded in the images that narrate the hunt, perhaps telling the hunters, not only where game could be found, but also when. We are still discovering how these cave paintings came to be and speculating about their significance. What is undeniable is their magnificence and poetic beauty, revealing to us a representation of a culture rich in symbolism dating from the Aurignacian era, some 40,000 years ago.

There is a symbiotic relationship between culture and technology, particularly in the realm of media communication. Technology historians and philosophers such as Lewis Mumford (1934), Jacques Ellul (1964), Elizabeth Eisenstein (2005) and Harrold Innis (1991) have written much on the subject. Lev Manovich writes in *The Language of New Media* (2001) that:

> As distribution of all forms of culture becomes computer-based, we are increasingly 'interfacing' to

> predominantly cultural data—texts, photographs, films, music, virtual environments. In short, we are no longer interfacing to a computer but to a culture encoded in digital form. (p. 69–70)

The technology contributions to our pattern language draw upon the concepts and works developed largely by Roland Barthes, Gunther Kress and Theo van Leeuwen, reframing them to fit the functions they help us identify. Because we are talking about the Web, we need to be mindful of those technologies that enable and/or constrain what we see. We also find that much of what we describe in terms of technology is also inextricably linked to design considerations, as the same constraints often lead to either design compromises or radical breakthroughs.

Often technological considerations stem from access and usability issues, especially when regulatory bodies such as the FCC and the European Union require Web designers and website owners to comply with certain infrastructure standards, thereby affecting, in some way, the final visual product. Web usability researchers such as Jakob Nielsen, Jared Spool, Steve Krug, Carolyn Snyder, and others, maintain blogs and websites that, in addition to the books they have published, provide the Web researcher with invaluable and timely information. Technology is moving so fast that we cannot always wait the full publishing cycle of books and proceedings to assess what is happening on the Web. Additionally, when we study the websites of other countries and cultures, we have to be able to distinguish between those elements we see that are enabled by technology, e.g., Flash™, and those that are constrained, e.g., connection speed. Absence of high-bandwidth technologies like Flash™ in a website may be the consequence of limited telecommunications infrastructure and might not necessarily indicate a lack of visual aesthetic sophistication in a particular culture. On the other hand, in some cultures 'less is more' and, regardless of the availability of high-bandwidth technology, those websites might exhibit an intentional simplicity.

Countries that don't have high-bandwidth constraints at the consumer level may be countries where individuals pay for all phone calls, including local calls. Paying for each minute on the Internet would influence user expectations regarding content. Some cultures rely heavily on Internet Cafés or other 'pay-per-view' facilities and non-entertainment based, or gratuitous animation would cause people to click and go elsewhere for their information.

In addition, as new technologies emerge, a technology/cultural design lifecycle is constructed. The first stage of a website's development is often based on mimesis—copying styles and layouts from the dominant technological culture until technical patterns are mastered. Only then do cultural aesthetic patterns reassert themselves. When Macromedia's Flash™ technology was relatively new to the world and only large multinational websites could afford the professional staff to build using Flash™, most of the websites we studied did not use it and there was a great deal of aesthetic diversity. As Flash™ technicians around the world became more proficient, more numerous and less expensive to hire, an aesthetic homogeneity began to show up, particularly in economically developed countries—sites displayed less cultural diversity as everyone tried to do the same things with Flash™ and a Flash™-built site became its own genre. Eventually, the cultural aesthetic began to re-emerge, cultural diversity in design reappeared—until the next design technology was launched.

The most influential Web innovation launched to date is Web 2.0: social media technologies like Facebook, Wikipedia, Flickr, LinkedIn and Pinterest. State-of-the-art websites cannot ignore the power of social media technology. However, there is very little scope for cultural aesthetics or localization in these sites. A Facebook site or page will look the same no matter where in the world the site is seen or the author is located. Every Facebook page has the same basic layout, color palette, orientation, and audience interaction. Website authoring and content management systems like WordPress.com and Tumblr.com also constrain the ability to design using cultural aesthetics and norms. A technological hegemony is established by the template design team. To be sure, a website's author may choose a template that culturally resonates with regard to color and spatial orientation, but subtle values regarding time modes and information flow, tempo and context are overridden by the dominant functionality of the underlying technology.

In essence there are two major considerations we need to pay attention to when looking at the technology aspect of our pattern language. General background knowledge on the state of technological infrastructure of the culture being studied is essential and a basic understanding of who pays how much for what is necessary to inform our research. More difficult to measure, because we can only do so by using comparisons, is the extent to which what we are seeing is influenced by a culture's aesthetic or by an underlying dominant technology feature or architecture. Let us go back to Alexander's *A Pattern Language* for a useful analogy. If we wish to build a two-story A-frame house, because it is the most efficient for our life-style and environment, we will find it difficult if the materials, blueprints (code) and the basic architecture available to us is for a three-story Victorian. In both technology and design, form follows function.

Technology patterns

- Audience interaction (Manovich), Chapter 7
- Information/Task complexity (Pettigrew), Chapter 5
- Information flow (Hofstede), Chapter 5
- Time mode (Hall), Chapter 6
- Time tempo (Levine), Chapter 6
- Uncertainty avoidance (Hofstede), Chapter 9

Design

"Design is the art of inventing and shaping two-, three-, and four-dimensional forms that are intended to satisfy needs, wants, and desires, thereby effecting changes in the attitudes, beliefs, and actions of others" (Buchanan, 1995, p. 48). The philosophy of practical design came into its own in the twentieth century with architecture, and in recent decades with product and graphic design. With this focus also emerged the practical application of aesthetic principles, once the purview of the fine arts. In the introduction to *Discovering Design: Explorations in Design Studies*, Richard Buchanan (1995) writes that prior to this shift in perspective there was a "… tendency to elevate special instances of production to the status of fine art while dismissing the vast range of useful and often beautiful production as merely the result of vocational or commercial enterprise, unsuitable for serious study." He goes on to note that, "The development of disciplines such as the history and philosophy of technology, material culture, and cultural studies signifies an effort to reassess the importance of things in cultural life" (1995, p. x).

No digital artifact is as culturally situated as a Website. The culture may not be a geo-ethnic culture: it could also be corporate (Starbucks), idealogical (The Sierra Club), or political (the U.K. Labour Party). All seek to provide information while at the same time conveying, through imagery, modality, and interaction, a representation of that organization's culture. The first job of the website's designer is to get a sense of that culture and what visually resonates with the culture's constituents (members?). As Bolter and Gromala write in *Windows and Mirrors* (2003):

> A Website is a digital artifact that is situated in an enormous variety of contexts…. To design a good Website or any digital application, the designer should try to understand as many of the contextual levels as possible: from the psychology of the individual user to the constraints imposed by the user's language, systems

> of beliefs, and popular cultural knowledge.... Cultural awareness should inform the design process from the beginning. (p. 139)

A website is shaped by the culture in which it is created as well as the culture for which it is being designed. In turn, these culturally situated designs impact the design of other environments within that culture—space ads in magazines, displays in shops, right down to the set background in television news studios and the layout of crawlers and boxes you see on the television screen.

The world of design is rich in pattern language. In the realm of color and color palettes, we look to theorist Josef Albers whose seminal work, *Interaction of Color* (1971), has been described as the only book one might need to understand color as a designer. My inclusion of Albers as a resource might seem to some as contradictory to my goal of establishing a pattern language. After all Albers maintains that, "Anyone who predicts the effect of colors proves that he has no experience with color" (interview,1968). Albers makes us mindful that all visual aspects are in some way relative and may be perceived differently in different contexts. Instead of using ill- or too broadly-defined terms such as *red* and *blue*, we should look for descriptors that tell us more about the color palettes favored by a given culture: muted, bright, saturated, dark, light, warm, cool. As Albers himself points out, while Western culture might teach us that red is a warm color and blue a cool one, there are cool reds and warm blues.

Our pattern language also calls on the expertise of designers Tim McCreight (1996), Robin Williams (2005), Lynda Weinman (lynda.com), and John Bowers (2008), for terms that describe design and aesthetic principles. These help us identify universalities and particularities in the websites we examine. Visual design concepts that transcend media forms also help us by providing a shared language. Lester Faigley, et al, outline key visual concepts in *Picturing Texts* (2004) that include "emphasis" (e.g boldface type), "point of view," "narration" (voice), "juxtaposition," and "metaphor."

In *Reading Images* (1996), Kress and van Leeuwen give us the term *modality* to describe content validity—which mode of content is more *believable*. In our pattern language we use content modality to distinguish between image and word for the information presented to the user. Some cultures prefer words and privilege them as conveyors of truth while others privilege images: "What is regarded as real depends on how reality is defined by a particular social group" (p. 163). Michael Ball, in his chapter "The Visual Availability of Culture" in *Image-based Research* (1998) draws particular attention to the fact that a culture's visual artifacts are instrumental for comparative analysis: "All cultures thus exhibit a rich visual character which is visually available" (p. 136). Modality is culturally based along three patterns: the image/word ratio, the use of white space (content density) and motion (still versus dynamic) in the visual plane (O'Connell, 2010).

Hofstede's 'uncertainty avoidance' is an example of an overlapping pattern term. It is a concept we use in discussing both technology and usability patterns. Similarly, we may use 'uncertainty avoidance' to assess the quantity of detailed information displayed on a website. Cultures with high uncertainty avoidance provide users with a great deal of information, often in multiple modes, with redundant navigation schemes.

Because I am, at heart, a usability engineer and cognition-based designer, in addition to the rich language that already describes patterns in the *art* of design, I include those patterns anchored in the *science* of design: those anchored in cognitive psychology, perception, and information processing. Visual processing happens at both the sensory and cultural levels. These are so intertwined that we seldom realize that what we are seeing is as much influenced by what we culturally expect to see as what our visual apparatus constructs for us to see.

Cognitive researchers such as Colin Ware (2008) and Stephen Kosslyn (2006, 2007) contribute terms that reflect concepts anchored in perception and visual processing. Mirroring refers to the brain's mirror neurons or cells whose behavior suggests we can mimetically evoke resonance with users. "Indeed, some scientists speculate that the mirror neuron system may be critical to the complex social structure maintained by humans" (Ware, 2008, p. 137) Deixis is a term that describes spatial references and, as I have previously stated, a website is all about 'where.' The deitic properties of buttons and links can be culturally influenced.

Two more important cognitive terms also create spatial metaphors for us to use. "Cognitive thread," according to Ware, "is the sequence of concepts that are held actively in visual and verbal working memories, together with the links between them" (p. 138). As most Web users are primarily information seekers, a site's ability to establish and maintain a cognitive thread is important to a greater or lesser degree depending on the culture. In this regard, cognitive thread may have implications for patterns involving context and chronicity or time mode. 'Framing' is related, but not identical to, 'point of view.' Point of view captures something of the voice of the narrator, the author of the website. Framing refers to the control of the information and therefore is equally connected to information flow, or what Hofstede called 'power distance.' What is captured and presented to the website visitor, and what is not, is a function of framing and becomes conscious or unconscious element in a culture's self-representation.

Design patterns

- Audience interaction (Manovich), Chapter 7
- Color palette (Albers), Chapter 4
- Content density (Bowers, Williams), Chapter 5
- Imagery: context (Hall), Chapter 5
- Information/Task complexity (Pettigrew), Chapter 5
- Information flow (Hofstede), Chapter 5
- Modality (Kress & van Leeuwen), Chapter 5
- Static/dynamic (Manovich), Chapter 5
- Time mode (Hall), Chapter 6
- Time tempo (Levine), Chapter 6
- Uncertainty avoidance (Hofstede), Chapter 9
- Visual/verbal ratios (Kress & van Leeuwen), Chapter 5

Bringing It Together

This chapter explored the unique relationship between culture, technology and design evidenced in digital media. To create our pattern language and define a set of pattern words or phrases that convey the whole ideas that exist in this relationship, I pulled from existing terminology and patterns in the fields of anthropology, visual communication, cognitive psychology, technology and design and categorize those patterns in terms of how they resonate with specific audiences based on their communication function.

Research at the intersection of technology, culture and design is now coming into academic discourse. A number of scholars (Wuertz 2005; Callahan, 2005; Ess & Sudweeks, 2005; Thorlacius, 2004 & 2007; and Segev, Ahituv & Barzilai-Nahon, 2007) have recently examined cultural trends in the Websphere. I am sure many more names can be added by the time this book goes to print. Regardless, research in this area is still relatively young and is complicated by the rate of change in the technologies on which the Web is dependent. Unlike books, magazines, film and even television, which can be easily recorded for study, the website meticulously studied today can be replaced in its entirety in a matter of minutes—one refreshed screen changes the playing field.

Unlike film or video, it is difficult to record a user's candid interaction on a website without highly specialized equipment.

Because Web artifacts have this ephemeral quality, we need to identify the underlying structures and develop ways to describe the principles at work in the websites we study. Having a pattern language that provides categories for comparison and analysis helps us introduce consistency in our method. However, it is necessary to reach out to other disciplines to put this language together—technology, cultural anthropology, cognitive psychology and practical design.

Summary

In this chapter, I have highlighted the many theorists, philosophers, researchers and practitioners whose work informs or provides material for the bricolage of our visual cultural pattern language. Above all we acknowledge the work of Alexander, Ishikawa and Silverstein in *A Pattern Language: Towns, Buildings, Construction* (1977). It is the pattern of instruction and exploration I seek to emulate in this work. However, the authors maintain that, as with any language:

> The patterns are very much alive and evolving. In fact, if you like, each pattern may be looked upon as a hypothesis like one of the hypotheses of science. In this sense, each pattern represents our current best guess as to what arrangement of the physical environment will work to solve the problem presented. (p. xv)

We hope, in our pattern language, to tease out the virtual cultural corollary and, in some small way, do for the Web what Alexander et al have done for the dirt world.

Notes

1 Robert Logan writes, "A media ecology approach connects all aspects of communication and informatics and embraces not only the study of media but also the study of technology and language and the interaction of these three domains all of which form an ecosystem" (2010, p. 35). I would suggest that another word for that ecosystem is "culture".

PART 2: PATTERN TERMS & METHODS

The analysis of visual patterns, and the language that is built around what is found, are as pervasive in the world as culture itself. Clifford Geertz writes in "Art as a Cultural System" (1976):

> But the craft approach to art talk is hardly confined to either the West or the modern age, as the elaborate theories of Indian musicology, Javanese choreography, Arabic versification, or Yoruba embossment remind us. Even the Australian aborigines, everybody's favorite example of primitive peoples, analyze their body designs and ground paintings into dozens of isolable and named formal elements, unit graphs in an iconic grammar of representation. (p. 1474)

In this section, I attempt to identify those "isolable and named formal elements" that are reference points for our pattern language by cataloging and exploring their descriptive terms. To analogize from Alexander, a 'walkway' may be a gravel path or brick paving. Each descriptor tells us something about the patterns utility and durability. Each pattern element in the chapters that follow list and define terms we can use to describe those elements as they relate to the aesthetics of the culture we are researching.

But first, to be able to use these patterns, one must decide on an approach, a method for gathering data and documenting what one finds. Just as a recipe outlines the ingredients one must use to successfully create even the simplest dish, the approach one takes and the method for documenting, comparing and analyzing the patterns observed in the various Websites collected, outlines what kind of study one is undertaking and where it must start. In Chapter 3, the reader is given two starting points and some general methods for collecting and organizing the data. The first method takes a deductive reasoning approach and relies on knowing who the target user is. The second is more inductive, starting with a collection of websites and attempting to identify, from what we see, both who the target users are and from that what their cultural aesthetics might be. As the data accumulates, one can move from one approach to the other depending on the over-arching question the research is trying to answer.

At the end of each chapter are exercises called "Putting Patterns Into Practice" that help the reader experience the pattern explored in the chapter. In Chapter 3, there are also two important tools discussed: a User Persona and the Cultural Aesthetic Grid. The first describes the target user, the second describes and documents the websites in the research. Creating the User Persona can be done at the beginning if the researcher has identified the target audience of the website. There may actually be more than one target audience. For instance, a site devoted to breast cancer might want to provide information to patients, their families, researchers, physicians, care givers and non-profit fundraisers. Designers of such a site would want to have a User Persona for each of those constituents so they could identify how the site should be structured and where the "voice" and language of the site should be pitched. Educators can have the User Persona created in conjunction with the "Day in the Life of My Cultural Twin" exercise (Chapter 8) as these inform and validate each other. A complete list of the "Putting Patterns Into Practice" exercises is included below:

List of Putting Patterns into Practice Exercises by Chapter

Chapter	Pattern	Exercise
3	Methods	*Cultural Aesthetic Hunt*
4	Color Palettes	*A Color Mood Board* *Home Page Montage*
5	Content: Modality, Ratio, Density, & Flow	*Working with a 'Thirds' Grid*
6	Time: Mode and Tempo	*Time Quadrants and Maps*
7	Audience Interaction, Information Density	*Affordance Testing*
8	Cultural Attitudes and Values	*A Day in the Life of My Cultural Twin* (User Persona)
9	Usability, Uncertainty and Ambiguity	*Basic Funnel Flowchart*

Chapter 3

Getting Started: Two Methods

As with many research projects, getting started presents a "which comes first, the chicken or the egg" conundrum. In our endeavor to identify the cultural aesthetic we can start from one of two points: 1) the visual evidence in the form of websites or 2) the target user community these websites hope to attract. As each informs the other, there may be a lot of moving the focus from user to the websites and back to user again. It very much depends on the researcher's question. Am I trying to design for a specific target culture? Then establishing and documenting some basic characteristics of users in that culture is a good starting point. It keeps the designer from falling victim to scope creep—a term use to describe a process by which a project evolves to try to please everyone and usually ends up pleasing no one. Am I trying to find out what is aesthetically happening in the ethno-webspace of a particular place? Am I undertaking some digital ethnography? Then starting with the websites and identifying the range of aesthetics exhibited there is the first step. It will be like walking into a village and looking around at the tools and the architecture before conversing with the inhabitants. It gives you something to 'talk' about.

What follows are suggestions outlining two methods for getting started. There may be some repetition, as both methods share an overall work pattern. However, the important point is that each method has its own specific starting questions. If one is starting with identifying the target user, the questions must be about the user. If one is starting with looking at a large collection of websites, the questions must help the researcher identify common functional patterns first before moving on to discovering the culturally aesthetic patterns. Therefore the initial questions will be about the website's purpose.

Once these beginning points have been established, we can move on to methods for collecting and reviewing the data and for documenting and indexing our findings. The tone of the text may become somewhat instructional from this point on—after all, this text is intended to be a 'how-to.' For instance, I have included an exercise to help you get started and a tool I call a Cultural Grid, the history of which and its use is explained toward the end of this section. The intention is purely to get you started, which I hope you find helpful.

Projects Starting with the Target User(s)

Every website needs to focus on three questions: 1) Who is the target user? 2) What does the user want to do? 3) What does the site need to provide so the user can accomplish his or her goals? (O'Connell, 2010) When we use the term *culture* as a characteristic of our target user, we may be referring to groups that are geo-ethnic (Irish), belong to a particular age group (millennials), life-style or ideology (eco-minded), or political affiliation. Any target user of a website might be part of one or more of these cultures and the dominant culture for that user will depend greatly on the purpose of the website itself.

Rarely is the researcher or designer the target user or audience of a website or other visual medium like film or advertising. So compelling is the impulse to identify with the user, though, that the designer or researcher needs to be continually reminded that there are important differences. Signs are often posted in the cubicles and team rooms of designers: "Remember: YOU are NOT the USER!" Helping to define the user, or audience, is a matter of collecting data. Before collecting examples of websites, it might be instructive to create a target audience/ user profile or *persona* (Cooper, 2004). Web and application developers have long used personas to help make key interface decisions and Alan Cooper's *The Inmates Are Running the Asylum: Why High Tech Products Drive Us Crazy and How to Restore the Sanity* is a long-titled but reasonably sized reference on design with a significant section devoted to creating user profiles.

Creating a 'persona' or dossier of the target user, or users if there is more than one, is not only helpful, it is essential. It consists of listing the demographic data of the ideal target user—age, sex, education, income, living environment, interests, purpose/motivation, etc.—in as much detail as possible. Key to making the persona tool work is to give the target user a name, e.g., Joe, Helen, Tiki, Nick. Some persona profiles or dossiers are pages long. The more specific the needs and desires of the target user, the more extensive the profile must be to be able to help the designer make tough decisions regarding site or application features. In the process of designing for the target user, the designer can ask, "Would Helen need/ want that?" It is critical to also outline the relationship the user has to the website or application. Asking the question, "Why is Helen coming to this site?" might uncover subtle things like whether the user will be a one-time visitor or a frequent user. If the website already exists, keyword search logs can help determine not only what the user came to the site to find, but also what the user *calls it.* Another place one can find user driven language is in the comment sections of any interactive website, its "help" or wiki pages, or in its associated social media sites.

If you are a designer, you will be gleaning what you can about your target user from a few other sites they might use so you can design something new for them. If you are a researcher, you will be looking for clues in existing websites, print media, advertising and film that reveal the user's cultural representation based on the dominant patterns observed over many websites. A web statistics site like Amazon's Alexa.com can be invaluable in finding websites by country (India) or category (News). The site often includes analytics that give you information on gender, education, browsing location (home, school or work), and visitors by country. For instance, eenadu.net, a news site with the tagline, "The heart and soul of Andhra Pradesh," is ranked 106th in India overall and 1,280th in global traffic. If you were hoping to identify a website targeted at southern India, the information about this site would fit your criteria.

Consider the persona profile of your targeted user as a living document. Start with the basic information and then keep adding to it as you notice specific information that helps you with the patterns you see. Some designers have artist renderings or cartoons of their target users included in the persona profile. Another useful tool is to go to Getty images and use keywords pulled from the persona profile and see what images come up. It's all grist for the mill.

Once you have your user's persona profile, you can begin to collect websites you believe have been created for and targeted at that user's culture. As you begin to collect these sites, you will want to examine the purpose of the sites, their strategic intent, their voice, and any technological constraints or contexts as outlined in the next section.

Projects Starting with the Website(s)

A well-designed website that successfully attracts its target users does so by meeting the expectations of both the user and the enterprise that paid for the site to be on the Web in the first place. By starting with the websites, we 'reverse engineer' our way into identifying the target user. However, the target user may not be immediately obvious from the web address, the domain name or even the various elements on the home page. Some websites masquerade as focused information sites when they are actually PPC, or Pay-per-click marketing sites. These sites target anyone who will click on links listed on their home page because the site owners get paid, sometimes only fractions of a penny, whenever anyone clicks. It has become a popular form of direct marketing because the structure and nature of the web makes it possible to track the clicks—where they came from and where they go. It could be argued that there is a specific aesthetic at work, albeit it not a culturally based one. Because the information on the site is keyword driven and the display is often automated, meaning that the same template is used regardless of what topic the site is supposed to cover, it would be difficult to ascertain if there is an actual culture being targeted. Unless you are specifically researching these sites and their aesthetics—and, believe me, someone is—they may be of little use to you. Being able to identify the dominant patterns of these PPC sites, however, will help you distinguish them from other, more useful sites.

Just as it is useful to create persona profiles of the target users and understand what it is they wish to accomplish (as outlined in the section above), it is also useful to determine the purpose of each website in your collection, its strategic intent and goals. There are three key questions for each of the websites you include in your research:

1) What is the website's strategic intent? What are strategic intents of the different sections or pages?

2) What is the voice of the website? Is it casual, formal? How is this expressed in the labels and content?

3) What are the appropriate speed/load constraints given what you know of the technological ecology of the culture you are studying? (O'Connell, 2010)

Establishing these characteristics of the website you are looking at helps you identify dominant themes and layouts and determine some of the characteristics, values and attitudes of the site's target user, leading, perhaps, to the cultural patterns of that community. From this point, knowing who the target user is becomes essential. A site designed to attract and communicate to audiences *outside* of its home culture will, necessarily, attempt to resonate with *that* culture instead of its own. Sites built for tourism are an excellent example of this. The patterns in these sites are dictated less by the originating geo-ethnic culture and more by the culture of tourism and travel.

After collecting sufficient data to see some distinctive website patterns emerge, the researcher can begin to create a target user 'persona' profile. Collecting the data is the most time consuming part of any research project. But without the adequately analyzed visual data of the website home pages and sub pages, we have no patterns to validate using other visual media.

Collecting the Data

Before we can begin to determine anything about the culture of our target audience, we need to gather data, in this case, screenshots of many, many websites we believe are targeting that user group. It is wise to collect not only the home page of each website but also any sub-pages that have different layouts, color schemes or information content variations. Using some sort of visual sorter or display helps us see connections and patterns much faster than piece-by-piece analysis. I have often found it useful to create a PowerPoint or Keynote deck (presentation) for each target audience I am researching. If I find a cataloging or database system that incorporates the visual, I will not hesitate to modify it for my purposes. In doing a semiotic analysis of old 78 rpm labels (O'Connell, 2010), I used a CD database system for the Mac called CDpedia because it displayed a grid of CD covers. I simply filled the CD cover field with scans of the 78 labels and added some custom data fields for various elements such as recording label, matrix #, artist, and so on. The result was an instant visual display of those record labels that matched my search criteria. The truth of the matter is that there are few, if any, specific research applications that neatly show us visually what we want. We must inevitably jury-rig what we have or can find to get what we need to see. Whatever application helps us see in aggregate what we have collected will help us see patterns.

Once you have a persona or two based on your target user, and you have begun to find visual material that has been designed for them, you might start with some websites that fit your persona's profile. If you are looking for Irish environmentalists (a particular sub-culture), you would start with websites using Ireland's country code extension, '.ie' and perhaps use some keywords that indicate interests in nature or ecology. As Ireland is a popular tourist destination, you would filter out any sites that overtly promoted tourism, although you might look at links from eco-tourism sites to see what might lurk there. A quick visit to Alexa.com might confirm your selection, or it might indicate that the majority of visitors to the site are not from Ireland at all. Once you have confirmed that your site appears to be designed for your target user, you can begin to analyze what you are seeing using the various tools outlined in each of the following chapters: a color palette poster or Pinterest page, a 'Thirds' grid, a navigational affordance test, etc. Each is explained in a section following the chapter called "Putting the Patterns into Practice."

We must keep in mind that some websites will have more than one target audience or may be directed at a particular subgroup (e.g., Irish millennials interested in eco-tourism.) We may need hundreds of screenshots to see reliable patterns. Whatever patterns emerge from fairly large data sets are worth noting and testing. We should also examine the sites we've selected for brand driven outliers—are these multinational brands or local brands? Remember, one is not looking to find a single exemplar but rather a pattern emerging from many examples. If we are studying geo-ethnic cultures, we must find sites that have a country's extension instead of .com (e.g., .ie, .de, .in), making every effort to find sites created and authored within that culture. For contrast, we should also collect some multinational sites, large brands like Coca-Cola or Kelloggs, but making sure these have the same national URL extension as the culture we are researching. In this way, we can be sure those sites are targeting that culture's population, although those companies may not have made any attempt to aesthetically localize the site's design.

Seeing and Validating the Patterns

Cultural patterns are often a product of environment. We should not be surprised if there are more patterns of space and cool colors in northern climes where snow and ice are part of the landscape much of the year. As the artist Millet put it in a response to comments on his ability to convey the essence of the farming countryside quoted in *Artists on Art* (1945), "My critics are men of taste and education, but I cannot put myself in their shoes, and as I have never seen anything but fields since I was born, I try to say as best I can what I saw and felt when I was at work" (Goldwater and Treves, pp. 292–293).

Once we have made a list of patterns that we see, we can ask questions. Do we see a pattern of white space? When we plot out the screenshots do we find there is a higher percentage of colorful areas? These, as you will learn in the succeeding chapters, will indicate a visual preference that could be culturally derived.

Once we have identified a pattern, we need to verify it. Two useful and readily available visual media that can be used are cinema and advertising, on television, in print and on the Web. In film, the cinematographer and director have spent a great deal of effort (and money) creating an aesthetic *ambiance* to convince the viewer of where they are in the film, regardless of where the film was actually shot. A dramatic example is the highly successful *Lord of the Rings* trilogy, shot in New Zealand but depicting J. R. R. Tolkien's imagined world called Middle Earth. Success for this film, and any other that relies on a sense of place, is the audience's suspension of disbelief. They don't expect us to notice this consciously: they expect us to accept the ambiance they have created and respond to it, have it resonate with us.

Television and space advertisements are also excellent for validating one's observations. They are tools of persuasion and rely heavily on audience reception and resonance. Ads have to speak to the target audience or they fail. They cannot afford to repel those whom they seek to persuade. Often, the aesthetic preferences of the product's targeted audience has been exhaustively researched. The increase in visual artifacts available on the Internet makes the job of collecting this evidence easier than ever before.

When I first began developing this pattern language in 2004, YouTube™ had not yet been invented. Web 2.0 was in its infancy as were social media sites and applications such as FaceBook™ and MySpace™. Twitter™ didn't come along until 2006. Macromedia's Flash™ (now an Adobe product) was not in widespread use and sites that did use it were easily identified by the "[---- Loading % ----]" message commonly displayed on their home pages. Consequently, most international websites, at that time, were still in basic, simple HTML. It was much easier to identify cultural aesthetic preferences back then.

In recent years, template driven content management systems (CMSs) like WordPress™ have become much more prevalent. There are also dominant 'functional' patterns that supersede the aesthetics of a particular culture. News sites that are modeled on print newspapers, for instance, will have similar layouts across cultures. To see the cultural differences expressed on these sites, one must examine elements such as the visual/verbal ratios, the content density and site navigation preferences (monochronic vs. polychromic) to see if, within the scope of the 'news site' pattern, there are differences that might be culturally derived. What visually tells someone from Portugal that this is a Portuguese news site and not a Brazilian or Mozambique news site? What visual patterns tell someone whose language is English, that a news site is American, British, Irish, or simply the English version of any other news site?

In short, anyone undertaking to use this pattern language approach must find examples of many sites, look at them together as a great mosaic, and let the patterns emerge. In the chapters that follow, I often mention the research that I, along with my students, have conducted using this method and various pattern-identifying analytical tools. The most dramatic example of a cultural aesthetic pattern emerging from this research was that of the northern European countries of Denmark and Sweden. We called the pattern 'Scandinavian Design' because that is what we saw clearly reflected across all the visual artifacts: websites, cinema, space ads, architecture and, of course, furniture design. Clean lines, ample use of white space to organize content, high level of functionality, a monochronic approach to navigation, and muted cool or natural colors were hallmarks of this aesthetic in over 100 of the collected websites. We first studied this pattern nearly a decade ago. It may still hold true today but that can only be determined by collecting examples of these websites as they are currently rendered.

Accordingly, I must emphasize that the examples I may use in the rest of this book are generally relevant for that *moment in time* when they were first researched. You may not be able to see that exact website any longer because a more current redesign has taken its place. A new Web technology may have come on the scene and its templates temporarily overtook or dominated what cultural aesthetic would normally be visible. However, as time has shown, the aesthetics of that culture will be evident somewhere and likely continue to be valid for the message creator who wishes to communicate in a deeply resonant way.

One more word of caution. We need to keep in mind that we ourselves bring to each viewing of cultural artifacts, the assumptions of our own culture. As art historian Michael Baxandall (1988) writes:

> [S]ome of the mental equipment a man orders his visual experience with is variable, and much of this variable equipment is culturally relative, in the sense of being determined by the society which has influenced his experience. Among these variables are categories with which he classified his visual stimuli, the knowledge he will use to supplement what his immediate vision gives him, and the attitude supplement gives he will adopt to the kind of artificial object seen. (p. 40)

In order to test whether we have managed to acknowledge and bracket our own personal assumptions about a culture's aesthetics and modes of self-representation, wherever and whenever possible, we need to find validation for our empirical observations through our networks of colleagues and friends who may themselves identify with the culture we are studying. If possible, we should verify our observations by tapping into the cultural groups themselves, perhaps with online surveys with mocked up examples of sites featuring key cultural patterns.

We should also go to the works of those who have written about these cultures to see if the patterns we have identified have been documented elsewhere. This is not to say that we will blindly accept contradictions to our findings but that we will see disjunctures between our evidence and their testimony as part of the 'thick description' of the pattern (Gilbert Ryle in Geertz, 1973, p. 6). According to Geertz, culture is "a multiplicity of complex conceptual structures, many of them superimposed upon or knotted into one another, which are at once strange, irregular, and inexplicit, and which he [the researcher] must contrive somehow first to grasp and then to render" (p. 10). In reference to color, Berlin and Kay (1969) point out that "semantic universals do exist in the domain of color vocabulary" and that "these universals appear to be related to the historical development of all languages in a way that can properly be termed evolutionary" (p. 1). The analysis we are doing with our pattern language is, in truth, ethnography and, as such, requires 'thick description' if it is to be truthful. It is also predicated on the assumption that there are some universals we can reliably count on to describe what we are seeing.

Finally, it helps to have a research practice for documenting all the observations you will make as you collect and examine the visual material you will use to discover the patterns embedded in these websites. Glaser and Strauss' Grounded Theory (GT) is a framework that is uniquely suitable for extracting meaning from the diverse materials collected. It helps you "understand... taken-for-granted assumptions and rules" (Charmaz, 2006, p. 21). The benefit of Grounded Theory is that, as patterns emerge in the data collection process, coding categories "grounded in the data" suggest themselves. For a fuller description of Grounded Theory, I refer you to Kathy Charmaz's excellent text, *Constructing Grounded Theory*, and Glaser and Strauss' (1967) seminal text on the subject, *The Discovery of Grounded Theory: Strategies for Qualitative Research.*

In order to avoid cognitive narrowing, it's best not to think too much about the early patterns, but simply to make note of them. This practice is referred to as 'memo-writing' in Grounded Theory. "Memo-writing forms a space and a place for exploration and discovery. You take the time to discover your ideas about what you have seen, heard, sensed, and coded" (pp. 81–82). For many years, I have practiced memo-writing in my Morning Pages, a discipline I learned from Mark Bryan's *The Artist's Way at Work*, wherein I handwrite three 8.5" by 11" pages first thing in the morning as a stream of consciousness meditation. It is not at all unusual for subconscious bursts of insight to tumble onto the page or for simple observations to organize themselves into recognizable patterns.

Using a Cultural Grid

Throughout the process of observing, reflecting and validating the data that has been collected, it helps to have a systematic way to categorize all the visual and transactional cultural cues. That is where a cultural grid helps us index our findings into the pattern language. I was first introduced to this grid in a course on Globalization/Localization studying under Bill Gribbons at Bentley University. As I began using Bill's grid with my students, we continued to shape and modify it to meet our own research needs, changing and further refining the categories which have now become the basis for our pattern language. An example of this grid is provided at the end of this section. The terms in the grid may be strange to you now, but each is explained in the chapters that follow. Next to each pattern variable is the chapter in which the pattern is defined and described. I strongly suggest that for each website you collect, you fill out this grid and include a screenshot image of the relevant pages of the site as well as the full URL.

This cannot be over-emphasized. When you find a site that speaks to your research, you must capture as many images of the site as you can and identify elements of the way the site works. Make sure you have a systematic file naming convention that will help you quickly find these images when you need them. There is no guarantee that the site you looked at on one day will be identical to the one you visit the day after—or even later that afternoon. It takes only a moment to completely change what shows up at any given URL address. If a website does change, the visual information and the differences between the two versions of the site provides evidence of something and is also worth noting. The old site is still valid as long as it spoke to the aesthetics of the target user's culture. Cultures do not change quite as quickly as websites do, so both the old and the new are data and provide material for the researcher to analyze. If, for some reason, you cannot find the image of the site you were analyzing, there is always *The Wayback Machine* Internet archive (http://archive.org/web/web.php). It doesn't save everything but you would be surprised at what you can find there. I have found the very first website I ever designed back in 1994, although the *Wayback Machine* could only go as far back as 1996.

The key to using the grid is to fill it in for each website just as you would assemble notes for each participant or interviewee in an ethnographic study. If you are skilled in databases, you can create a form to collect this information that can then be searched according to the common terms you use in your descriptions. For that reason, it's best to be consistent in your terminology. Whatever makes sense to you is fine, as long you as you stick with it throughout. In Chapter 10, I include some analytical examples that demonstrate how that can be done. However, my own methods evolve as new needs crop up. Feel free to adapt the grid to your specific research needs. It is merely a tool and can always be improved.

A cultural grid, such as the one at the end of this chapter, serves as a pattern key or an index of the observations the researcher/designer has made for each of the websites (or other visual media) created for the target user. Just as the persona profile tells us about the user, the cultural grid tells us about the aesthetics of the user's culture.

Summary

In this chapter we outline two approaches for getting started with your research. The first starts with what we know about the target user. The second uses the evidence provided by the websites themselves. This chapter also discusses data collection and recording methods necessary and suggests a Grounded Theory approach for recognizing emergent patterns.

Immediately following this chapter is a cultural aesthetic grid for use in documenting observations for each website. The grid itself is based on a similar tool used in a Globalization/Localization course taught by Bill Gribbons at Bentley University that has been extensively modified to include the pattern language criteria outlined and discussed in this book. Also following this chapter is an exercise to help you get started on a search for a culture's aesthetics. It involves observations of all forms of visual media and is the preamble to the other activities and exercises found at the ends of each of the remaining chapters.

Cultural Aesthetic Hunt

Purpose: To practice finding useful websites and to begin 'seeing' what's hiding in plain sight.

Method:

1. Collect 10 examples of websites from your target culture that fit into the following categories (you may have to look at five to ten times that many narrow the list):
   - 3 multinational sites that have been localized
   - 3 national sites
   - 3 local sites (serving a local community or sub-culture)
2. 1 wildcard site
3. View 3 to 5 films produced and directed by and for your target culture.
4. Collect 10 examples of ads or print material (web, magazines, newspapers) from your target audience.

You will have to be careful to distinguish when these visual media are speaking to 'insiders' and when they are targeting 'outsiders' such as tourists. Also, while we want to identify the aesthetics of 'cultures' we must use countries as a starting point as so much of the media we need to look at is categorized according to country, e.g., the country code extensions on URLs like .de, .ie, .fr, .uk, etc. In addition film databases tend to list the country in which the film was produced. Be on the lookout for subcultures within your target country. You may decide to focus on a subculture or you may focus on the culture 'clash' that is growing in your target country.

For the websites: To ascertain whether a multinational site has been localized, compare it to the '.com' version and other country code extensions for cultures that you believe may be quite different in their aesthetics. For all URL extensions, if there are multiple language options (often indicated by little flag icons in the upper corners or in the menu bar), check to see if the visual aesthetic differs from one language option to another. If you are using Google's Chrome browser, use Google Translate to see if the copy is consistent across all the language options. Having a language option at all may indicate that the site is either targeted at several subcultures or at 'outsiders' or tourists.

For the films: Try to determine what the director and cinematographer are doing to give you a sense of place. Free write several pages of your impressions.

For the ads and print material: Look at layout and imagery and make notes on your impressions. Remember, space ads in magazines are intentionally visually dominant. This should not suggest that a culture whose web evidence indicates a verbal modality has necessarily switched to a visual preference because more than 50% of the ad's space is graphic. An ad that is even 25% verbal content is privileging the word.

Compare your notes to what you are seeing on the websites and make notes on your cultural grid regarding your impressions.

Table 3.1 Cultural Aesthetic Grid

Pattern Variable	Site Title: Web URL
Color Palette (Ch. 4) Warm/cool Saturated/Muted National/Natural Contextual/Brand	
Content (Ch. 5) Modality (Image/Text) Ratio (% Visual, % Verbal) Density (High/Low)	
Information Flow (Ch. 5) Fast/Slow Poly-dimensional/Uni-dimensional	
Time – mode (Ch. 6) Polychronic/Monochronic	
Time – tempo (Ch. 6) Fast/Slow	
Audience Interaction (Ch. 7) Static/Dynamic Serial/Branched/Hypermedia Passive/Interactive	
Information density (Ch. 7) High/Low	
Imagery (Ch. 8) Collectivist/Individualist (many people/one or two persons)	
Context (Ch. 8) High/Low	
Attitude (Ch. 8) Rational/Emotional Objective/Relational	
Value Orientation (Ch. 8) Materialistic/Quality of Life	
Usability (Ch. 9) Ambiguity tolerant/intolerant	
Uncertainty Avoidance (Ch. 9) High/Low	

Chapter 4

Color Palettes

> *Every culture on the planet lives in a world whose colors have assumed powerful metaphorical significance.*
>
> —Lynell Burmark, *Visual Literacy*

You may notice that this chapter is labeled color (or colour) palettes and not simply color. This is because all websites use color. Color is light and it is light which lets us see. Even black (the absence of colored light waves) and white (all the light waves combined), are essential to seeing and showing. Whether it is 'light through,' (the web, film and television) or 'light on' (paper and other reflective media), black on white is the color combination that provides the greatest basic contrast and so, as with the page you are now reading, is the preferred color palette for verbal text, with white on black a close second. All other combinations provide less contrast than black on white.

As a pattern, **color palette** is often a subtle indicator. While the hue itself (the pure color as determined by wave length) may not be a particular cultural preference, other attributes of color may be. It is far more likely that patterns concerning color will be evident in color properties other than hue: saturation or intensity, temperature (warmth/ coolness), and symbolism or association.

A Color Primer

Before we appropriate any color terms for our pattern language, it might be helpful to review how and when these words are normally used. Readers who are well versed in the subject might skip over this and go straight to the significance of these words in terms of our pattern language.

All color is seen because of light and its wave length properties. A rainbow is the phenomenon of light passing through and being refracted by water molecules into the full visible light spectrum from red, the longest wavelength, to violet, the shortest. At each end of the spectrum are light waves we cannot see, infrared and ultraviolet rays. In the absence of light, there is no color. I think this is most dramatically illustrated in the moments before the sun comes up and the world appears to be awash in shades of gray. There are two light modes and related color systems: 'light on,' which describes the properties of reflective light and uses subtractive color, and 'light through,' which describes the properties of projected or filtered light and uses additive color. "In subtractive color, all light waves except those containing the color we see are absorbed or 'subtracted' by a surface" (Bowers, 2008, p. 90). These 'light on' colors are created through pigments (paintings) or offset printing (brochures, magazines, printed material).

Additive color is created directly by light passing through a television, film projector, computer monitor or digital device. When all the colors in an additive color system are combined, the result is white or pure light. A common measuring system for color as it relates to 'light through' technologies is HSV (hue, saturation and value) or HSB (hue, saturation, brightness.) *Hue* is the primary attribute of a color. It is the property that makes the color discernible as red or green or blue (RGB: the three transmitted colors that, in combination, produce all the other colors we see on a computer screen or television.) How deeply red or blue is a factor of its *saturation* or how intense the color is and is relatively measured as a range between a color's highest saturation and its least, or gray, where the closer to gray it gets the more muted the color is. *Brightness* measures how light (tint) or dark (shade) the color is or its *value*. When we are talking about color pigments on paper or canvas, we make a color a darker shade when we add black. To give it a lighter tint, we add white. In 'light through' systems, to darken a color one removes some of the light and to lighten the tint, we add more light. Black is the absence of light, and white is the presence of all the visible light wave spectrum. So hue (the purest single wavelength for a color), its saturation and brightness or value are all ways to accurately measure a color. But we can also talk about a color's coolness or warmth when a particular color is combined with another.

Color *temperature* is something that is not apparent to everyone. In the blue color family, we can have, ice blue, teal, sky blue, azure, aqua, midnight, lapis and cobalt. Midnight and cobalt describe cool blues, whereas teal, azure, and aqua describe warmer blues. Notice how the color names relate more to environmental characteristics or objects than something measurable like bright blue or pale blue. There is an underlying assumption that we can refer to the environmental object in order to convey a color's properties. However, I doubt one could recognize the referenced blue if one saw raw cobalt—although cobalt has been used to produce the color blue since the Bronze Age—or raw lapis lazuli, which, despite its rather purple hue, was the basis for painting the sky and the robes of the Virgin Mary in Medieval and Renaissance times.

Color labeling based on perception is highly relative to the viewer. And yet we can perceptually experience the effect of a color's temperature, its warmth or coolness, in terms of foreground and background. Warm colors tend to advance to the foreground. Cool colors tend to recede, creating a perceptual hierarchy that disappears when the color is removed leaving a gray-scale image. The important thing to keep in mind is that any color can be described as warm or cool depending on the colors in close proximity and the colors that are added to the base color. A bit of green or yellow will warm a blue. A bit of blue will cool a red.

How does this translate into cultural aesthetics? For example, certain cultures, and I generally notice this in terms of climate, prefer a warm or cool palette. Of the sites researched between 2004 and 2012[1], those that were north of the temperate zone tended to prefer cool colors, especially in their blues. Those that were in the more temperate-to-tropical zones chose warmer colors. Using blue again as an example, Mediterranean countries like Greece and Italy generally chose blues that were a bit greener or warmer. At the same time, I noticed that Italian sites, while using warm blues, also tended to use other warm colors and there was often a preference for strong vibrant colors. This seemed to also be the case for other Latin cultures such as Spain. Similarly, Latino cultures in the New World also preferred warm, vibrant or highly saturated colors.

How these colors—indeed, all colors—are perceived when next to each other is also property of a color's interaction. "Color deceives continually.... One and the same color evokes innumerable readings" (Albers, 1968, p. 1). This is why any color palette pattern requires looking at a multitude of examples together to identify dominant aesthetic patterns.

Color and Culture

A 1969 study by Brent Berlin and Paul Kay investigated the number and variety of color words collected by native-speaking informants in nearly 100 languages using a set of 329 color chips. The prevailing view at the time, as proposed by Edward Sapir and B. L. Whorf, was that "language is semantically arbitrary and relative to every other language" and that "the search for semantic universals [patterns] is fruitless in principle" (p. 2). Berlin and Kay's hypothesis was that the ease with which color words translated suggested that Sapir and Whorf's doctrine of linguistic relativity was not valid, at least not for color words. They found that there was consistency to the pattern of words that 'named' basic colors and their variations. Their study launched whole new areas of study in linguistic and cultural perception and is worthy of further investigation by the reader. For our purposes, it is most important to note that the primary distinction in color terminology was between bright/warm and dark/cool colors.

In our final property, color symbolism or association, color can be associated with national colors, religious symbols, cultural practices as well as environmental attributes: the colors of leaves, mountains, water, sand and rock. Victoria Finlay's *Color: A Natural History of the Palette* (2004), opens with a description of ochre, the first color paint. Ochre, or iron-oxide, can be found in all parts of the world, from Siena in Tuscany to Newfoundland in North America to the outback of Australia. Today, we think of it as a yellow to brown color but it can also be red, white, even blue. For the Aborigine, it is not just a pigment—ochre is sacred, the art of its use secret and reserved to those who are specially trained.

Evaluating the Color Palette

In Figure 4.1, we see a table of home page screenshots taken from a single domain extension '.it' for Italy. To get this montage, I took screenshots of each website and placed each screenshot on a slide in Keynote™. PowerPoint™ works just as well. I then used the slide sorter view to do a quick grouping of dominant color palettes, the cool colors at the top and the warmer colors toward the bottom. The overall effect is as

Figure 4.1 Home Page montage
Home Page screen shots of 30 sites with country extension '.it.'

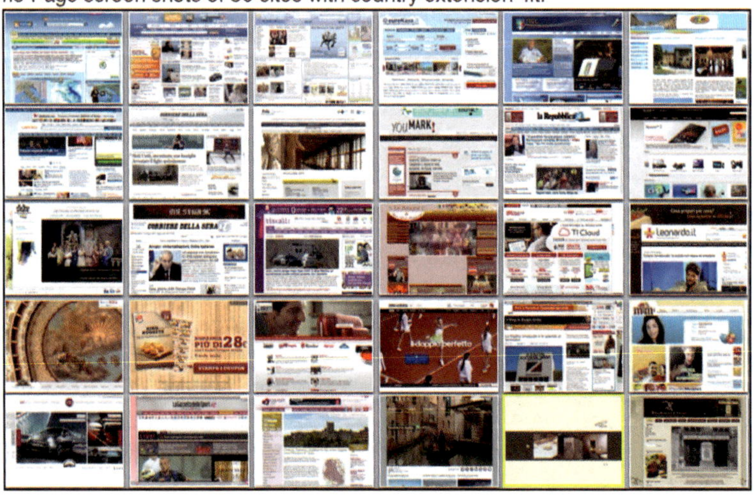

if you had a tray of colored tiles and moved them around until you had a suitable arrangement. This example has only thirty screenshots, but many more can be used and zooming in and out lets you adjust the level of granularity you want to use to view patterns. Squinting your eyes to blur the image sometimes helps.

Once I had a good arrangement of commonalities, I took a screenshot of the slide sorter window and brought that image into Photoshop to remove the extra gray spacing between the 'tiles.' This was done by slicing the rows of images in the screenshot and then nudging the rows up close under each other. I then saved the image to use here. Obviously, using more screenshots would increase the validity of what we are seeing. This example is merely to show you the technique.

Of the images collected only two are multinational sites: Burgerking.it and Adidas.it. Only one site displayed significant white space—a local artisan bakery site in the bottom right corner. Indeed, a cursory examination of this collection suggests that these sites, and perhaps this culture, prefer intense warm colors. Even the blues tend toward warm 'Mediterranean' blues, the few cool blue spots being informational photography and not decoration or navigation. In informational photography, the properties of the colors cannot be assumed to be there by design, and therefore not as reliable as data for cultural aesthetics.

As we look for palette trends, we need to be aware of several issues. It should not surprise anyone if a McDonald's or Coca-Cola website sports a significant amount of red. Indeed, biologically, red and yellow are high response colors, a point that does not escape most successful multinational brands. But highly saturated versions of these colors may not appeal to all cultures. In an effort to adjust for localization, occasionally, a site targeted at a culture that prefers muted colors or significant negative space will reserve its use of bold brand colors for the header and use white or a light neutral color for the body of the page. Websites for governments or national sports teams—think soccer or *fútbol* —will prominently use the colors of the national flag. If your collection of sites includes a great many of these nationalistic websites, observations based on your montage could be skewed. I have generally found it useful to divide sites into the following categories, keeping in mind that the cultural target audience for all categories should be roughly the same: 1) multinational, 2) national (government or national entities), and 3) local services. It might also be useful to collect images of sites targeted at outsiders to see if there is a difference in that culture's self-representation.

If the culture studied has a multitude of sites devoted to a particular function, for instance 'news,' a collection of news media sites may be instructional. These sites might be in direct competition with each other and, hence, might be particularly sensitive to culturally based aesthetic considerations. Indeed, the heavy concentration of a particular site genre or function—news, media entertainment, gaming—may also be examined for cultural referents because, in good design, form follows function. We never know until we begin to look at them together.

In addition to the symbolism of national brands, cultures have other, symbolic references regarding color (Cabarga, 2001; Hardin and Maffi, 1997; Peterson, 2000; Smith, Moriarty, Barbatsis & Kenney, 2005). While these don't always apply to what you are seeing, it certainly is wise to research cultural symbolic meanings regarding certain colors. In Asian cultures, particularly China and India, red is considered a very lucky color and not the color of danger often cited in Western culture. Chinese New Year's gifts come in red envelopes. Red garments are worn by brides where in Western cultures they wear white. White in some Asian cultures is associated with death and mourning. A most interesting, and painstakingly researched, graphic "Colours in Culture" by David McCandless (2009, p. 76) can be found at http://www.informationisbeautiful.net/visualizations/colours-in-cultures/. This color wheel coordinates 13 colors with 84 symbolic meanings spanning ten geo-ethnic cultures. Red, according to McCandless' wheel symbolizes for the Chinese happiness, fertility, marriage and success. However, information of this kind should be used to validate the evidence that one already sees on the Web and not to essentialize the color palette of a given culture.

As you fill out the cultural grid (Chapter 3, Table 3.1) for each website you collect, make note of general tendencies and outliers in terms of the color palettes you are seeing and look for possible theories for what you are seeing.

Summary

This chapter provided a primer on how color works and explained why various color palettes appear to dominate the mediascape of various cultures. The particular characteristics of color we use in our pattern language stem from basic color science and we carry those terms forward to help us create subtle distinctions between one aesthetic preference and another. Of course, there are colors that show up in nearly every website. This, too, can be explained, according to Josef Albers, by the interactive properties of color. The presence of color language is, as Berlin and Kaye have demonstrated, a universal phenomenon.

We also examined how we can identify the color palette of a culture we are researching. I provided one example here as a home page montage using the slide sorter of a standard presentation application, but there are also examples of other types of comparisons in Chapter 10: Using the Pattern Language. By noticing certain trends and patterns, communicators can identify not just the colors that resonate with an audience but also their intensity, temperature and symbolism.

Notes

[1] I have been teaching the course *Visual Media in a Cultural Context* since 2004 resulting in examinations of over 2000 website instances across 30 countries that span the globe. Websites from the same company examined in different years are considered different website instances because of changes to the website in the intervening period. As the intent of the course is for the students to develop this different way of "seeing" a culture's aesthetic, collecting all the data into a larger study has never really been considered. However, that doesn't negate the findings which appear to hold up, year after year as I allow the students to discover these patterns on their own.

A Color Mood Board

Purpose: To get an overall impression of the color palette(s) for your target culture

A mood board (also known as swatch board or sample board) is a visual technique for pulling together a collection of disparate pieces of information that capture the design 'mood' or aesthetic. Web designers borrowed this technique from interior designers to help them elicit from their clients a more focused fleshing out of design ideas including colors, typography, imagery and patterns. This technique can be done using traditional collage (poster board, paste and images torn out from magazines) or digital media and social media apps like Pinterest and Olioboard. An excellent tutorial for color mood boards can be found at http://www.creatively-driven.com/the-art-of-the-mood-board/

Method: Start by brainstorming some adjectives that will help frame the style for your board using the pattern terms from the Color Palette pattern language, e.g., dark, muted, bright, clean, soft, etc.

If you are using traditional collage approach, rapidly go through 10 to 20 magazines tearing out anything you see that reflects the colors, textures and imagery you have observed in your cultural aesthetic hunt. I have found international travel magazines like *National Geographic* and *Condé Nast* have the most potential material. After 10 or 15 minutes of collecting idea examples, assemble and paste your 'samples' on a poster board.

If you wish to use digital media, you can copy and paste images into a Powerpoint™ or Prezi™ presentation. You can also use social media applications like a Pinterest™ board. The advantage to using Pinterest is that you can quickly and easily create different boards for different aspects of your target cultures aesthetics, and move or copy elements from one board to another.

Each method or approach has its benefits. I often prefer doing things by hand for their meditative and synaesthetic benefits. However, a traditional collage may then need to be photographed to include in one's notes. Mood boards that are created digitally suffer from being on too small a 'canvas.' On the other hand, they are more easily shared and edited.

Both methods can be used at different points in the research project giving you the best of both worlds.

Home Page Montage

Purpose: To step back and 'see' the dominant color palette used in a culture's sites

Method: This technique is outlined at the end of this chapter on color palettes. It requires collecting screenshots of home pages from websites that have been designed for the target culture. The more screenshots you collect, the better the results will be. It is important to keep in mind that you are not looking for a specific color but rather attributes of the colors you are seeing on these sites. There are many blues and reds. But are they bright or muted? Warm or cool? The distinctions may be subtle. The benefit to using the slide sorter on a presentation application is that you can move things around until a pattern emerges—not unlike one of those tile puzzles where you slide the tiles around until the pattern is revealed.

There may be some cultures that prefer or avoid certain colors. This, too, is important and may emerge from the montage. It may also be the case that some colors are used, but only in a particular context. The colors of a country's flag may dominate government sites and sites promoting national sports teams. This does not necessarily indicate a 'cultural' preference or significance for those colors.

This technique can also be used to compare overall visual/verbal modality and content density. Often, looking at just the individual home pages prevents us from "seeing the forest for the trees"—each individual home page, as a tree, might lead us to believe there is incredible diversity in a culture's webscape. It is not until we can see the "forest" of home pages that we can discern patterns worth noting.

Chapter 5

Content: Modality, Ratio, Density & Flow

The unity of form and content is, where it occurs and to the degree it occurs, a cultural achievement, not a philosophical tautology.
—Clifford Geertz, "Art as a Cultural System"

When we talk about a website's *content*, we are dealing specifically with a site's visual and verbal elements and their relationship to each other. In traditional visual content analysis we seek to select, code and quantitatively analyze a collection of images, e.g., ads for Mattel's Barbie doll or images of war in the news (Rose 2001, 2007, 2012). However, when we are looking at the Web, we engage in a more complex milieu. In our pattern language, we want to look at content in terms of its modality, function, state of permanence, and visual/verbal ratio. We also look at its spatial attributes: juxtaposition and content density. We examine navigational content for its flow, tempo and subtle indicators of a culture's power relationships. Some of these are more measurable than others. All of these taken individually and together give us many clues to a culture's aesthetic. No single content aspect of a website is something that can be discretely measured. It interacts with several other measures both denotative and connotative, such as static/dynamic interaction, imagery context, attitudes, and values. The first step in examining those measures is an extensive visual content analysis. Therefore, this chapter outlines the use of an analytical grid and the content information we can learn from it.

A website's information content falls into two major groups: 'dynamic,' the informational aspect of content that can change from day to day, even minute to minute; and 'constant,' i.e., that which is structural, decorative or navigational. In some Web pattern languages, the term 'static' has been used for the decorative/navigational content. However, in today's multimodal digital environment with context driven interfaces, navigational elements may change based on the dynamic content of the page. Decorative content may change in color and imagery to help the user recognize that they are in a different area of the website. By using the term 'constant' we are indicating that these elements remain structurally the same from day to day, but not necessarily from page to page within a single website.

Both groups of content can be coded and measured using three elements: modality, visual/verbal ratio, and content density. In addition, navigation can be evaluated in terms of time mode, time tempo. Various access applications, as well as navigational links may indicate power relationships.

We also want to identify which content pieces are decorative, navigational and informational. Informational content is the most likely to change. Some informational content is 'editorial' and some is 'advertising.' Advertising generally refers to visual or verbal material that is not 'in-house', i.e., promoting the brand of the site itself but is, instead, 'paid placement' of information that helps to monetize the site. In-house methods for monetizing an information site is often accomplished by selling subscriptions to its content, as do many news and specialized information sites, like NYTimes.com and ConsumerReports.org. Decorative content, on the other hand, is, or should be, part of the branding or identity of the site and generally include logos and typography unique to the website's owners/authors. Navigational content consists of menus, buttons and constant or persistent links to other parts of the website.

Using a 'Thirds' Grid

To code and measure these elements, the use of a grid superimposed on a screenshot of the webpage allows us to better gauge percentages and layout patterns. My preference is to use a 'thirds' grid (Figure 5.1). This is a compositional tool commonly used in photography and videography for locating the best 'points of interest' in one's image. In this grid arrangement, the plane of the image is divided into three sections both horizontally and vertically, yielding nine equally sized parts. In photographs and video, where the vertical and horizontal lines intersect are 'points of interest,' locations within the plane that the human eye finds more balanced and aesthetically interesting. Digital SLR cameras at both the professional and consumer level have embedded thirds grids the photographer can turn on and use to compose the image before shooting the picture. By placing the subject of the image at one or more of these intersections, the photographer works with the way a viewer sees, rendering the image more aesthetically powerful. Even the most horrific images achieve a certain aesthetic beauty and power when the composition uses the rule of thirds.

Figure 5.1 Rule of Thirds Grid

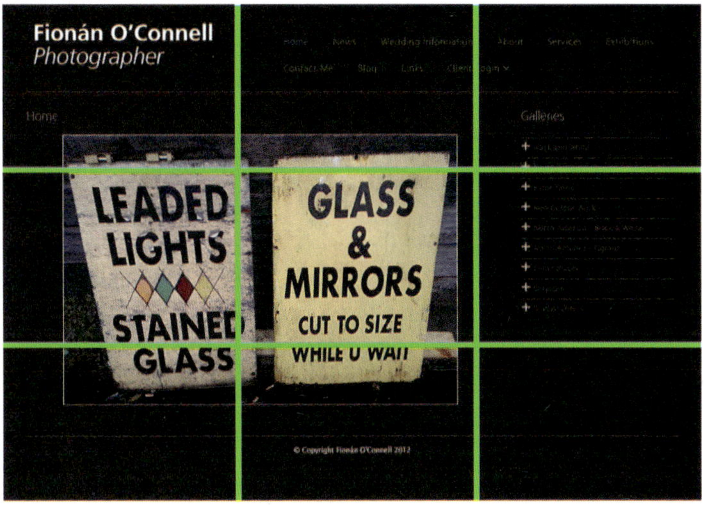

A simple 9-cell grid helps us measure the visual/verbal ratio and identify the function of each visual or verbal element. The left and center cells in the middle row are 100% visual. The top left cell is decorative/branding, the top center and entire left column are navigation.

Figure 5.2 Z-pattern scanning

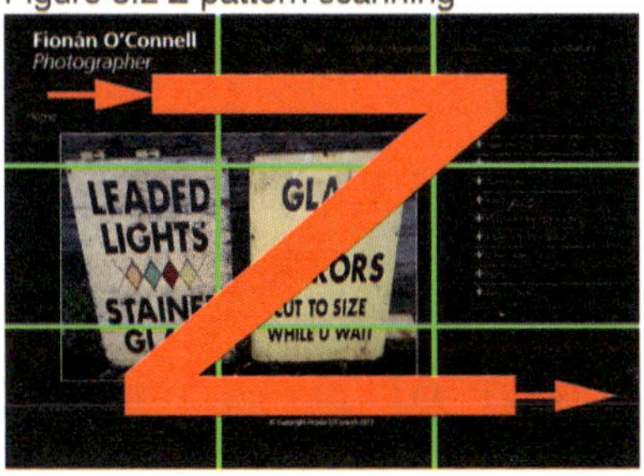

Figure 5.3 F-pattern scanning

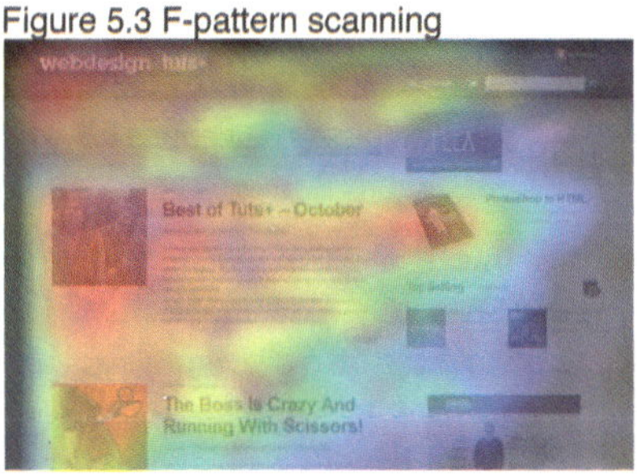

We can also use this grid to reverse engineer the conscious or unconscious composition decisions of photographers, videographers and designers by looking at how they have placed important visual elements in the plane of the overall image. Researchers looking for image placement in a particular genre of website, e.g., news sites, can use a thirds grid to identify the structural design of a website that may change the content from day to day but maintain the placement of the verbal or visual elements.

Another reason for using a thirds grid exploits what we know about people's eye search pattern. Eye tracking studies I conducted as a usability engineer have shown that there are two dominant scan patterns used by website visitors: the Z-pattern and the F-pattern. Z-pattern (Figure 5.2) scanning suggests that the most important information on a website should be placed along the top row in our thirds grid. Placement of content at the next level of importance should be where the lines of the Z meet with the intersections of the horizontal and vertical thirds. The 'call to action' should be located in the bottom right corner. Usability tests where the 'call to action' button was placed in the bottom left-most corner and the 'cancel' or 'reset' button placed in the bottom right-most corner, at least in Western user groups, confirm that users will click on the button in the bottom right corner, usually without even reading the label. The user's expectation that this would be the location of the 'call to action' button or link overrode the actual words on the button.

F-pattern scanning (Figure 5.3) appears to be a pattern preferred by users of information heavy sites (Nielsen, 2006). The scanning is not unlike the skimming of traditional media like newspapers, books and magazines, where the first line is scanned to see if this is the target information being sought. One hypothesis for the different patterns may be linked to content modality: users that use Z-scanning strategies showing a preference for image modality and those that use an F-scanning strategy preferring websites that have a high verbal ratio. Another hypothesis might be that it depends greatly on the purpose of the website itself. Information websites that scroll two or more screen lengths may elicit F-pattern scanning strategies, while websites that are single screened with a prominent call to action will promote Z-pattern behaviors.

Whether these patterns exist because the users have become acclimatized to the layouts or because web designers kept moving elements around the page until they finally hit a successful pattern is a chicken or egg question, and perhaps worthy of investigation. Whichever came first, we now have these two patterns that most designers choose to use. It is also important to recognize that these patterns became established when the technical dominance on the web was largely Western and content was predominantly in English. Cultures that read from right to left, e.g., Arabic, or from top to bottom, e.g., Chinese, did not have the means of displaying those cultural interaction modes because early HTML did not allow for, or accommodate, either the structural layouts or typesets that reflected those cultures (Barber and Badre, 1998). As the underlying source code was itself written in English and had left-right reading direction, it follows that everything that was displayed in the first years of the Web stemmed from those norms.

We can perhaps hypothesize that in cultures that read right-to-left, the Z- and F-patterns will be reversed or mirrored. The important thing to remember is that people, no matter where they are or where they come from, do not read everything on a website in a linear fashion. The medium itself appears to encourage scanning over reading. Users are generally in an information 'search' mode when they visit a website, seeking what Jared Spool (2004) calls the 'scent of information,' the subtle cues that tell a user that they are likely to find what they are looking for if they click on that button or link.

Modality

Just as content falls into two groups, constant or dynamic, Web content falls into two visual modes: images and words. Gunther Kress (2010) uses the term 'mode' to describe a "socially shaped and culturally given semiotic resources for making meaning" (p. 79) People tend to have a preference for one content mode over the other and that appears to be culturally consistent, all other things being equal. A website's 'modality' indicates which information mode is privileged for that culture, which tends to be more present and, consequently, relied on as the conveyor of information. In this case 'privileged' means: Which modality is the one more easily 'believed'? Which is taken as 'truth'? Some cultures believe what they see—some, what they read. In addition, according to Kress (2010), "Societies have modal preferences: *this* mode is used for these purposes, *that* other mode for those other purposes.... If a mode is preferred for a specific social domain it will become more specialized in that respect" (p. 83).

Modes include more than just visual words and images. They also encompass spatial elements such as layout and dynamic, i.e., still or moving. An additional mode not present in early websites, but now ubiquitous as a result of YouTube™, is the aural mode, as in music or speech. While some websites are truly multimodal, containing all manner of modes, most are composed of words and images and these can be coded and measured to give us a sense of a culture's dominant modality and spatial aesthetic.

Visual/Verbal Ratios

Using a thirds grid, the researcher can estimate what percentage of the visual and verbal is contained in each of the nine tiles. This is called the Visual/Verbal ratio and is an indicator of a cultural preference for believing what one *reads* (words) or what one *sees* (images) (Kress and Van Leeuwen, 1996). As always, the purpose of the site must be first taken into account. Even in the most verbally based culture, a photographer's website will have a preponderance of images. It is best to begin evaluation with sites where, all things being equal, both modes are informational and relevant, such as a news site (Figure 5.4). The number and size of images on the front pages of a news site may reveal a good deal about the aesthetics of its target audience. The cultural reliability, meaning how much can we rely on the target audience being of that culture, is also very high because, while we can assume that some news is of international interest, the vast majority of the news is for the people closest to where it is happening.

Figure 5.4 Rule of Thirds Grid with four different news websites: Iran, Argentina, Denmark, India.

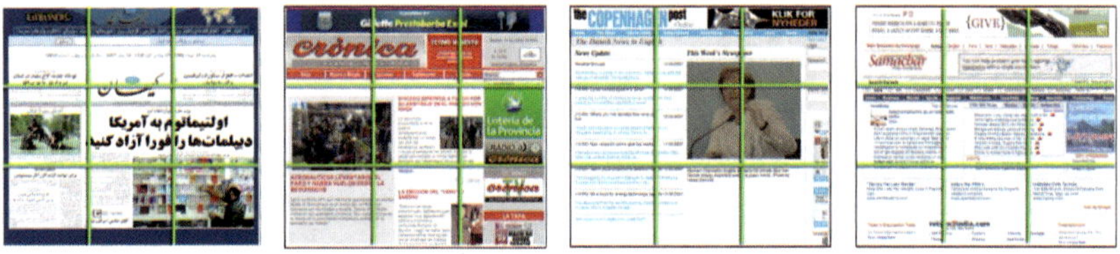

Note the heavier use of color and imagery in the two left-most pages and the higher percentage of verbal material in the two right-most pages. The *Copenhagen Post* has the most white space.

The visual to verbal ratio is read at several levels. It helps us refine our evaluation of a culture's modality by measuring to what extent a culture prefers one modality over another. Again, the thirds grid is essential for determining the visual/verbal ratio.

The first pass should be of the overall webpage. Leaving typeface aside, estimating what percentage of the content in a section of the grid is a matter of identifying how much of the space is occupied by words and how much is not. Intentional negative space should be considered visual if, in the researcher's opinion, it is warranted. For this reason, it is best to start with a uniform aspect ratio dimension. Any aspect ratio based on the screen size will work (see Figure 5.5) as long as one is using it uniformly: 640x480, 800x600, 1024x768, etc. It helps to create the grid as a layer in PhotoShop, or some other visual editing application, to the specification common to all the websites in the study.

Figure 5.5 Establishing a standard screenshot size

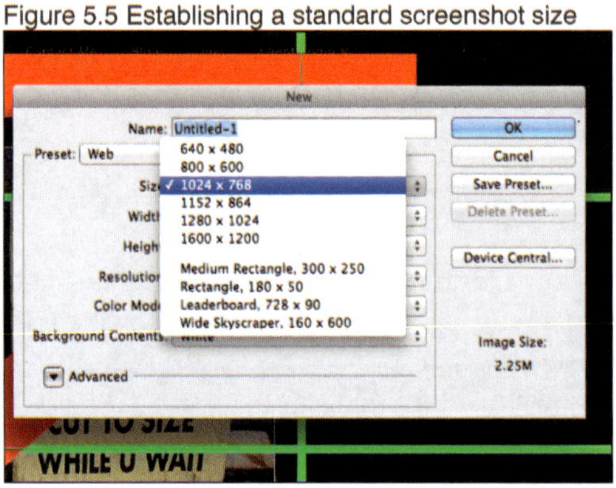

That way the researcher can normalize what is seen and those sites that have a smaller width will register negative space to the right and left of the content areas. This is a subtle clue regarding a preference for display constraints. This is accomplished by setting the browser window so that it will allow screenshots of the same size as have been set for the grid.

Once it has been determined how much actual verbal content is in a cell, the ratio estimate should be adjusted to allow for spacing. Titles get more spacing before and after than do paragraph lines. Whatever rule of thumb is used, it is important that the observations are denotatively consistent across all the websites being examining. The spacing issue works the same for the visual content as there will, naturally, be some spacing between images in the information content areas.

After completing a visual/verbal ratio assessment across all nine cells, the researcher can add up the data and compute an overall ratio. Once that is done, we can begin to look at the content in terms of its function and do a similar assessment for each function telling us whether the culture prefers one modality over another consistently across functions.

Content Density

Another important aspect of content is the layout in which it appears. Content density describes how much 'white' or 'negative' space there is. High density means there is little white space while low density indicates that there is a good deal of white space. Two of E.T. Hall's cultural dimensions appear to be closely related to content density: chronicity and context. These are explained more fully in chapters on Time (Chapter 6) and Cultural Attitudes and Values (Chapter 8).

Cultures that prefer high density layouts might also be low context (everything explicit), polychronic (multi-tasking, random order) and desire high levels of interactivity. A typical Facebook.com layout would be the result of these preferences. Not surprisingly, the design of the Facebook page comes from a low context, high-tech culture that encourages multitasking behaviors to fulfill a monochronic (linear) strategy, i.e., gather demographic data on users in order to monetize their interactions (Gorlick, 2009).

Alternatively, users from cultures that prefer low density layouts find high density layouts confusing, too busy, and irritating. They need the white space to help them visually organize the content and to provide a sense of order. They might also be low-context, but are more likely to be monochronic. Or they could be high-context, monochronic. The zen-like low density qualities of traditional Japanese design is very different from the efficient and organized low density preferences of northern European cultures. When we look at the traditional architecture and arts of Japan, we see that white space is not empty, but is instead full of restful, uncluttered space. The navigation may be high context and enigmatic to outsiders, simplified to levels of extreme simplicity. But to those in that culture, this level of simplicity is expected.

The only multinational website that has ever achieved this level of simplicity is Google—a true exemplar of negative space for both the art and the science of simple functional design. However, Google too, encountered some common pattern expectation problems with their early, mostly Western users. At first, users did not interact with the site, waiting a patient 45 seconds for the rest of the page to download, not realizing that there just wasn't anything more. As Google's Marissa Mayer explains in a 2012 interview concerning Google's home page design with Bianca Bosker of the Huffington Post, "I'd ask them, 'What are you waiting for? … They'd say, 'I'm waiting for the rest of it.' The blank home page was so out of context in 1999 that they were just waiting for the rest of it." Google's solution was itself quite simple. They put a small copyright notice at the bottom of the page signaling that it was okay to begin searching the web. To this day, the only embellishments Google uses on its home page are the artistic variations on their logo to celebrate various events and holidays.

Function

In our content analysis, we need to determine which function the image or text fulfills: decoration, navigation or information. Decoration and navigation will be relatively constant and consistent over time—the expense and impact of changing one's visual identity on the Web, one's brand, colors and logo, is never undertaken lightly. Informational content will change, perhaps daily. For this reason we need to look at the balance or ratio of visual to verbal information in terms of their function and dedicated space. If a site's content is more than 50% visual, but those visuals are decorative or provide symbolic navigation, perhaps because of language concerns, then one cannot argue that its modality is significantly image-based because the information is actually provided by the words.

Karen Schriver (1997), in *Dynamics in Document Design: Creating Texts for Readers*, describes five ways that images and words can be combined for effective design. While her work does not specifically reference cultural attitudes towards one combination or another, adding her analysis of these combinations gives us more ways to incorporate rich description in what we see.

The most common arrangements of integrated words and images include those that are redundant, complementary, supplementary, juxtapositional or stage-setting. Each arrangement describes the relationship between the modes and the message. For instance, in redundant arrangements, both the words and the pictures carry the same information and the same level of information. In complementary arrangements, one of the modes, either the words of the image provide *part* of the message. Both modes are required to understand the message.

Supplementary arrangements in websites might provide the best clue to a culture's modality. If the main part of the message is in the verbal mode, then even when there is an accompanying image, the modality of that message is verbal. If the main part of the message or the key information is the image and the words provide just a bit more information, we can hypothesize that the culture has a visual modality—it believes what it sees. For this reason, juxtapositional arrangements may be poor indicators of modality but excellent indicators of other cultural properties. In juxtapositional arrangements, both modes are used but there is tension between them, usually witty or humorous. Clever advertising is often modally juxtapositional. And humor and wit is deeply cultural and very high context, even in traditionally low context cultures. Jokes do not easily translate as there are embedded cultural values and attitudes that create the humor.

The fifth arrangement of modes, and perhaps the most common seen in websites, is stage-setting. Every decorative element in a home page is concerned with providing visual background to the user so that they recognize where the are, and what they can do and find there. While it is possible to set the stage with words, it is often best achieved, just as it is with a real stage, with imagery. This extends to a greater or lesser degree to the typeface of the decorative and navigational verbal elements on the website. When we look at and think about typeface, we are not primarily concerned with the words themselves but with the artistic aspects of the way the words look. So typeface is a visual, not verbal mode and can do the work of stage-setting by subtly indicating values, ambiance, and tone.

Decoration

Visual or verbal elements that are purely decorative, outside of a website's logo and tagline, are almost a thing of the past. In the early days of the Web, it was thought that one needed to provide entertaining, decorative, images and animation to attract visitors and keep them coming back to one's site. That was back when 'surfing' the Internet was a novelty. As we approached the second decade of the Web, the reasons why people came to a website changed. Instead of Internet 'surfers' we became information 'hunters'—hence Spool's concept of the 'scent of information.' Decoration, unless it is in some way identifying where you are and what you can find there, has no purpose.

However, decoration that identifies has a great deal to tell us about the cultural patterns we are seeing on a webpage. When the color palette is dominated by the brand colors, one can speculate that there is some cultural serendipity there, e.g., Coca-Cola's red is perfect for places like China and India where the color red symbolizes good fortune. However, for the purposes of understanding content function, it is enough to know what is purely decorative and what is decorative in terms of identity.

Navigation

Navigation is central to the function of the Web. There are two modes of navigation and our expectations of what will happen when we click are set by what we have experienced in our Web interactions. Linked words and terms and menus *bring us somewhere.* Buttons *do something* or execute a function. For instance, a menu link will take me to another part of the website. A word or phrase that is colored, highlighted and underlined will bring me to either another part of the website or another website on that topic. A 'Subscribe' or 'Submit' button generally makes something happen. If that something is of consequence, e.g., it sends money to someone, or transmits personal information, it may even generate another screen with another button to confirm that we do, indeed, wish to proceed with this action. As described in *Visualizing the Web* (O'Connell, 2010), designers can test their users' expectations of these interactions with affordance testing. "*Affordance* refers to the perceived and actual properties of a thing, primarily those fundamental properties that determine just how the thing could be used.... Doorknobs turn, switches flip, and underlined colored text in a Webpage should bring you somewhere when you click on it" (p. 132).

But images that are not buttons can also be considered 'links' in cultures that privilege images over words. Menu items might contain or be entirely composed of icons. Links to information content might be presented as photographs or even images of logos or highly stylized words with the accompanying words providing supplementary information.

Information

Informational content needs to be distinguished by the editorial/ advertising split. Advertising is information and does not refer to the site's own branding. It is advertising when a site reserves areas for content that sells other goods and services. Paying close attention to the modality of advertising elements can tell us a great deal about a culture's aesthetic preferences. No other form of media is as researched and measured, each click a traceable response. The space devoted to advertising on a site is a precious commodity and direct response testing has developed into a fine art. Of course, getting the numbers behind these ads is difficult for an outsider. But if the research you are doing is on a website of your own or that of your enterprise, then you have a great deal of information available in your click-tracking database.

Hall and Hall, in *Hidden Differences: Doing Business with the Japanese* (1987), write, "Members of a culture not only share information, they share methods of coding, storing, and retrieving that information" (p. xvii). This insight contributes to our understanding of information flow —how fast or slow messages move within a culture—and with concepts of closure or completion. Information flow also concerns the path of information. Is it poly-dimensional: bottom-up driven and moving unrestricted through the websphere? Or is it uni-dimensional: top-down and controlled sharing? And how do we know?

According to Hall and Hall, information messages can be fast or slow depending on the speed with which the coded message becomes decoded. The logo on a T-shirt is a fast message while an entire outfit of clothing may take more time to decode. A news site's headline is a faster message than an in depth article. Both of these are faster than an e-book. An information graphic is faster than a work of art, whose symbolism requires a great deal of contextual knowledge. However, symbolic references familiar to a given culture are quickly decoded within that culture. Early Christians used the icthys, a symbol of a fish drawn with two curving lines. However, in another culture, this drawing could be decoded as an all-seeing eye, albeit without the pupil. Most important to consider in the design or analysis of websites is that, "fast messages sent to people who are geared to a slow format will usually miss the target" (1987, p. 5).

In the context of websites, information can be coded or embedded in both the words and images on every webpage. Images are, by their very nature, more ambiguous: rapid signifiers to those within a culture and slower and more culturally difficult to decode by outsiders. Outsiders must first determine if the images are decorative, informational or navigational. The more abstract the image, the more effort involved in decoding. Words, on the other hand, can be much faster to decode—if they can be taken at face value. One must also determine if the words on a translated page have the same message and meaning as those on the same page in the site's native language. The amount of effort invested in researching the speed of the message depends, of course, on the purpose of the research.

Information flow can also be described as poly-dimensional (bottom-up, free-flowing), or uni-dimensional (top-down, restricted or controlled).

Poly-dimensional flow cultures are often also high-context cultures, as with Japanese or Latino cultures where information spreads rapidly because of the nature of people's personal or organizational relationships. In Japan, people who work together spend a great deal of professional and personal time together in close proximity and in open work environments. Information is readily shared in the moment with little emphasis on "need to know" or critical information paths (Hall and Hall, 1987). Contrast that environment with American or Western European attitudes towards information, where it is often compartmentalized, focused, and ranked in terms of importance, all behaviors that tend to restrict or impede information flow.

Where these cultural preferences regarding information flow become interesting is in how the communication function of a website differs from face-to-face conversation. Despite efforts to make a website appear to be like a conversation between the website's owner and its user, as is often the case with large multinational corporate websites that exploit every conceivable method of social media and marketing, words and images on a screen are not the same as conversations between people. Geert Hofstede's (1997) concept of power distance is perhaps the closest measure of poly- or uni-dimensional information flow. The greater the power distance, the greater the tendency toward a top-down uni-dimensional flow of information and the greater the effort to restrict or control what information is disseminated and to whom. The Wikileaks.org website operates at one end of the information flow continuum. It states on its website that its goal is to:

> ...bring important news and information to the public. We provide an innovative, secure and anonymous way for sources to leak information to our journalists (our electronic drop box). One of our most important activities is to publish original source material alongside our news stories so readers and historians alike can see evidence of the truth. (http://wikileaks.org/About.html)

At the other end of the spectrum is any website that requires secure access and grants various levels of access permissions, either through the role of the person using the site (administrator, contributor, subscriber), or through some 'pay-to-view' arrangement as with subscription-based newspaper and magazine sites (Figure 5.6). Granted, these sites are difficult to research unless access is granted, either through a free-trial period or by special permission of the website owner. However, restriction-by-design must be factored into any assessment of the user's experience of both interaction and information flow. This section on navigation covers only the visual or modal aspect of navigation—what it looks like. How navigation behaves and how that can be coded for patterns is explored in Chapters 6 and 7.

Figure 5.6 WSJ subscription based "pay-to-view" keys

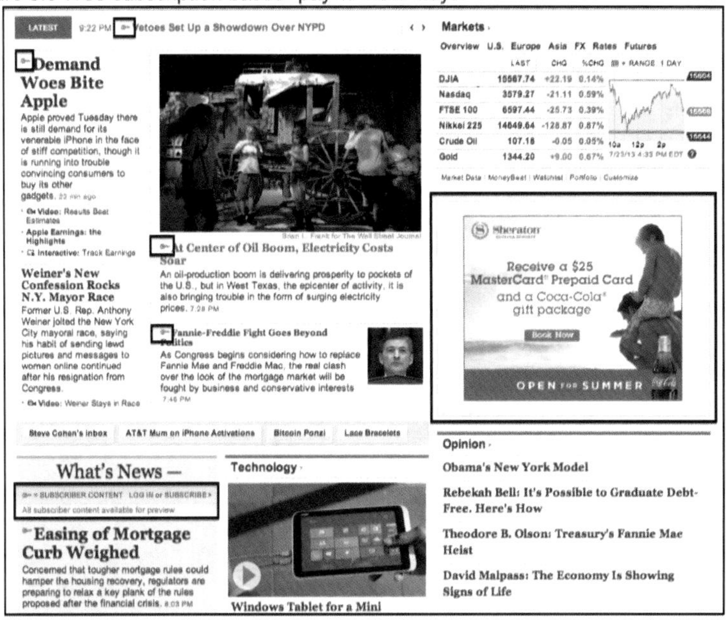

Other Content Clues

Patterns that reveal cultural attitudes and values are perhaps most informed by the work of Edward Hall and Geert Hofstede and are discussed more fully in Chapter 8. From Hall we borrow the concepts of *culture context*, as well as time and space. High-context cultures tend to convey information by what is *not* said or printed. Low-context cultures are more explicit and will not only provide a great deal of information but also ensure that access to that information is equally overt and explicit.

Information complexity refers to the amount and type of information needed to operate confidently in a task domain. It is related to the Gestalt principle of closure in that too little or too much can be anxiety producing. Cultures with minimal information complexity can tolerate 'just-in-time' information flows. Cultures with maximal information complexity demand contextual/historical/status information related to the source of the information (confidence/competence) and how it fits into the 'big picture'.

Hofstede's concept of power distance translates into *information flow* and *information access*. The flow of information and levels of control can be detected in the way a website works. Does this culture prefer higher levels of external controls? Or does it rely on internal controls?

Summary

In this chapter we looked at both the shape of the information situated on a webpage as well as the content of that information. We reviewed the use of a Thirds grid as a way to divide the space on the webpage to evaluate various ratios, including the visual/verbal distribution of content. We covered content density and the important function that white or negative space plays for some cultures whose aesthetic relies on that space 'full of nothing' to organize the layout and flow of information. In addition, we covered one of the most important elements of any website, its navigation.

While clues to the cultural attitudes toward content may be connected to modality and, indeed, have been referenced in this chapter, they are more fully explored in the following chapters.

Working with a 'Thirds' Grid

Purpose: To aid in computing mode ratios and functions in a webpage.

Method: If you are familiar with Photoshop™, you can create a thirds grid by creating a transparent layer with two vertical and two horizontal lines that carve the space into thirds, yielding nine cells or blocks. By placing that layer above all the other layers you can use the one transparent layer over and over again to do your 'quantitative' analysis of the use of space in each cell of the grid.

To do this using a Powerpoint™ or Keynote™ presentation application, on a blank slide (I use the black style or theme) use the shape tool to create a neon green vertical line. Copy and paste that line and then move them into position, dividing the plane into vertical thirds. Repeat using horizontal lines. Once you have the four lines evenly placed, select them all and 'group' them into one object. You will then be able to copy and paste this object over any slides you create in the presentation deck. There may be a way to create the same effect with the 'guide' lines, but I have not been able to figure it out and this method works just as well and can be copied and pasted into new decks whenever necessary.

Use a Thirds Grid to analyze the following:

- Percentage of image material per cell (modality)
- Percentage of verbal material per cell (modality)
- Percentage of navigational, informational or decorative material per cell (function/purpose)
- Percentage of negative space in a cell (density)

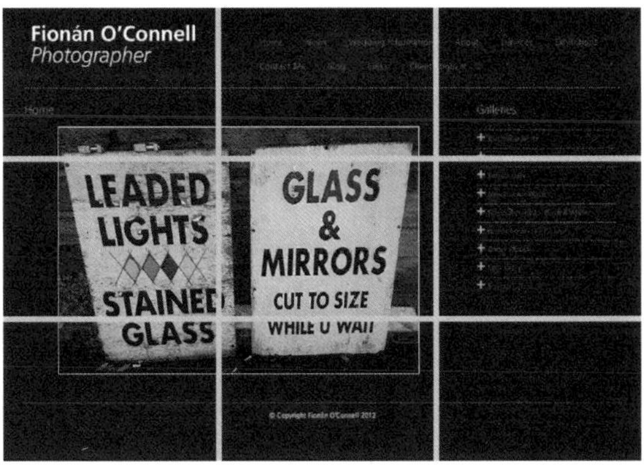

Chapter 6

Time: Mode & Tempo

> *What then is time? Is there any short and easy answer to that? Who can put the answer into words and even see it in his mind? Yet what commoner or more familiar word do we use in speech than time? Obviously when we use it, we know what we mean, just as when we hear another use it, we know what he means. What then is time? If no one asks me, I know; if I want to explain it to a questioner, I do not know.*
>
> —St. Augustine of Hippo

Time is perhaps the most universal and fundamental element of our existence. Time is not visible, nor is it tangible—and yet we are aware of time, how little or how much we have, and whether we have used it well or wasted it. It is a non-renewable, non-recyclable resource. Ever since the sundial was invented, a great deal of effort has been invested in mediating time in order to measure and manage it. Until recently, this has involved creating two- and three-dimensional artifacts like the hourglass, the calendar and the clock-face. The advent of the computer has allowed for more precise measuring of time and has significantly changed our relationship to it—at least in those cultures where technology dominates. Our relationship to time has generated a small industry: clockmakers, watchmakers, digital devices of every conceivable size and shape, consultants and programs designed to help us better manage our time, and when that fails, to help us deal with the stress of having no time.

Use of time and attitudes towards time are two important cultural concepts introduced by E. T. Hall in *The Dance of Life: The Other Dimension of Time*. "Time is a core system of cultural, social, and personal life," he writes in the Introduction. "In fact, nothing occurs except in some kind of time frame. A complicating factor in intercultural relations is that each culture has its own time frames in which the patterns are unique" (1983, p. 3). A culture's relationship to time, and the artifacts and explanations a culture creates in an effort to measure and manage time, are fundamental to how cultures understand self and identity. At certain points in our lives we experience time as stretching endlessly before us. At other times we feel that time, or rather the lack of it, is taking over our lives, causing us to feel stressed and out of control.

There are philosophical and visceral connections between time, control, stress and mental health that some cultures experience more than others. Is it coincidence that Western cultures, with their emphasis and dependence on advanced technologies experience more time-related stress than other, more naturalistic cultures? Does our relationship to time change as a result of new technologies or do we invent new technologies to reflect or express changes in our relationship to time?

Users in a given culture may also be constrained by technology limitations such as bandwidth capacity, network speed, or economic factors, e.g., 'pay per view' or 'pay per minute,' that affect their attitudes toward time as it relates to technology. Cultures where Internet Cafés are the dominant access point may have users that prefer to keep their highly active, processing heavy, interaction modes (covered in Chapter 7) reserved for game playing and will favor low or passive interaction modes for information seeking. In looking at a culture's relationship to time, it is perhaps best to leave issues of technology aside and focus on the sociological patterns observable off the Web. Sociologically biased time behavior patterns can be characterized as two separate, but related, concepts: time *mode* and time *tempo*.

Time Mode

Our perception of time is quite complicated and culturally derived. At one extreme we have Western, modern culture—a Cartesian construct of absolutist/reductive/positivist philosophy—experiencing time as mechanical, measurable, spendable. At the other extreme, Eastern cultures and Zen philosophy experience time as *being*, as a flow, no beginning, no end. Quantum or modern physics gives us a clue to this seemingly irreconcilable contradiction. Time is both. When we are so absorbed in a creative activity that we lose all 'sense' of time, it is flow. When time is present for us, i.e., when we are abundantly aware that we have too much or too little of it, it is a Heisenbergian[1] observed object. Neither state is sustainable for long. Just as we can discuss the modality in the physical senses of sight and hearing, we can look at these time states as modes.

Time mode relates to how a culture perceives and manages time. At one end of the spectrum is the preference for punctuality, schedules, and the need to tightly manage those schedules. At the other end is a preference for placing the importance of the task at hand and personal and professional relationships first, putting attention to the schedule a distant second.

Of course, not every individual in a particular culture will share these preferences. But it is interesting to note that guides to conducting business in other cultures, such as *When Cultures Collide* (Lewis, 1999), often cite respect for time mode preferences as key to successful business relationships. For instance, under the business practices for Mexico the authors make these observations:

> Punctuality is not high on the list of Mexican priorities. In most hierarchical societies it is accepted that the powerful make others wait. Easy access to the presence of a superior casts doubt on his status. In Mexico, life is not organized around a clock—a mere machine—but around a succession of encounters and relationships that are qualified and quantified not in linear fashion but in terms of depth of personal involvement, excitement, opportunity or caprice…. Mexicans recognise that fate has many things in store for us that we cannot foresee, that the sequence of events in God's calendar may not correspond to human schedules and deliveries. Sometimes one has to make changes to accommodate changes elsewhere—one has to choose one's priorities. (p. 424)

Contrast this with his observations for Austria: "In the west of Austria people are punctual and hate time wasting." As with neighboring Germany, time and the organization of it is central to their culture. Lack of punctuality is impolite and suggests deficiencies in other areas. However, Lewis also observes that time behavior in the capital, Vienna, is slightly different. "The 'coffee house' culture of the capital encourages leisurely gatherings where gossip and networking thrive and clock watching takes a back seat" (p. 211) Clearly 'time,' the one thing we all share from the moment we come into the world, is not treated in the same manner across all cultures. Time, in an abstract sense, is interesting. However 'Time' in a cultural sense is key to understanding relationships within a culture and between cultures.

In *The Dance of Life*, Hall noted that, "complex societies organize time in at least two different ways: events scheduled as separate items—one thing at a time—as in Northern Europe, or following the Mediterranean model of involvement in several things at once" (p. 45). He separated time *mode* patterns, and the cultures that exhibited them, into two groups: monochronic (M-time) or polychronic (P-time) cultures. Monochronic cultures tend to be 'one-track,' linear, doing one thing at a time. Behaviors and tasks are scheduled, compartmentalized. The schedule takes priority over other considerations. Timeliness, or staying on schedule is valued. Polychronic cultures, on the other hand, tend to be 'multi-track,' circular, oblique, allowing for many things to happen simultaneously. The task at hand or the relationship takes priority.

Over decades of research in all parts of the world, Hall made observations as to how cultures invariably fell into one of these modes or the other. Of particular interest is the fundamental relationships to time these cultures exhibit. P-time cultures appear to be inside time, oblivious to it. Relationships take precedent: "Polychronic people are so deeply immersed in each other's business that they feel a compulsion to keep in touch" (p. 50). In contrast, M-time cultures objectify time. Hall suggests that we pay particular attention to the metaphors used in M-time cultures to talk about time: "M-time is also tangible; we speak of it as being saved, spent, wasted, lost, made up, crawling, killed, and running out" (p. 48). When monochronic people find themselves in a polychronic mode they often will say that they were "lost in thought" or "lost all sense of time." There is a suggestion in those phrases that this behavior is not normal for them.

For Hall, monochronic time is arbitrary, a learned cultural preference. "Because it is so thoroughly learned and so thoroughly integrated into our [Western/American] culture, it is treated as though it were the only natural and logical way of organizing life. Yet," he emphatically maintains, "it is *not* inherent in man's biological rhythms or his creative drives, nor is it existential in nature" (p. 49).

In my own research I have come to observe that there are 'instance' people, those who are acutely aware of what time it is and exactly when they have to do things, and 'flow' people, who seem to be oblivious to time when engaged in a creative task that interests them. As research in the area of creativity bears out, many creatives, regardless of their geo-ethnic culture, observe polychronic attitudes and practices. Those who study creativity have come to see the sensation of flow, not as a mode of *time* but as a mode *living*. Mihaly Csikszentmihalyi has spent his career studying creative flow and what he calls "optimal experience." He categorizes cultures as those that promote or inhibit flow. "Cultures are defensive constructions against chaos, designed to reduce the impact of randomness on experience. They are adaptive responses, just as feathers are for birds and fur is for mammals" (1991, p. 81). Time-bound cultures inhibit flow. Hall would agree, maintaining that P-time cultures have more *flow* in them than M-time cultures do.

You might ask why all this talk of creativity and flow is relevant to the concerns of aesthetic patterns on the web. When we think about who designs the majority of the websites out there, we have to concede that the designer's personal time mode, her workplace culture and larger community culture together influence the outcome. There may be subtle aspects of a website's patterns attributable to the preferences of the designer's cultural aesthetic environment—and attendant relationships to time.

In *A Geography of Time*, Robert Levine (1997) gives us different metaphoric terms for P-time and M-time, referring to them as *event time* and *clock time*. "The difference between clock time and event time is more than a difference in speed, although life certainly does tend to be faster for people on clock time" (p. 82). He writes, "Under clock time, the hour on the timepiece governs the beginning and ending of activities," regardless of whether a goal has been achieved or a relationship nurtured. "When event time predominates, scheduling is determined by activities," the time allotted being open-ended (p. 85). An event is not an instance, it is the beginning of a flow as the boundaries of time are not as measurable as they are with clock-time, which comes pre-segmented as seconds, minutes, hours, days, months, and years. Time experience seems to fall into a dichotomy where the monochronic, clock-time instance is something that is objectified when it is noticed—and the polychronic, event focused flow is a state of being that is not conscious of objectified time.

In our modern technologized and hypermediated culture, we are increasingly experiencing time in this dichotomy. The Facebook minute is a good example. Defined as, "an elongated and obscure period of time spent distracted on Facebook when the original intent was to merely check your messages,"[2] time spent on Facebook, or on many other similar social media sites, passes without notice. However, at any website, once we hit the return key or mouseclick, our expectation of the response time for task completion is measured in microseconds. Delays in network response time tend to result in distracting, and often self-defeating behaviors. We are waiting for a webpage to refresh. It isn't doing that fast enough, so we open another browser window to do something else, e.g., check our email or our Twitter account. Perhaps we go back to that original window and click again. What we've actually done is increase the load the network has to process, minutely slowing the response time for everything. And yet, we feel compelled to keep clicking away—we call it multi-tasking.

'Multi-tasking'—attending to more than one task at a time—appears, on the surface, to be a polychronic phenomenon. However, polychronic cultures excel at juggling the many relationships and tasks that are the hallmarks of this time mode. It is not at all clear that the kind of digital multi-tasking currently overtaking our lives is functionally the same.

Recent studies of students who claim to be highly efficient multi-taskers are revealing that their perception is an illusion. A group of Stanford University students who maintained they could efficiently do many things at the same time, talk on the phone, send instant messages, and read their email, in reality performed very poorly in lab tests designed to measure multi-tasking. Not only were they less capable of doing any of their tasks well, they were also training themselves to be distractible (Ophir, Nass & Wagner, 2009). Given what we know about P-time and M-time, however, the question must be asked, were they less able to successfully manage multiple tasks because of the demands of the technology? Or was it because they are situated in an M-time culture and what we tend to measure are M-time activities?

Time Tempo

> *At its simplest, then, time is life and the events that give it shape, form, identity and meaning . . . The smallest indivisible unit of this time, this internally experienced time as life, is neither the minute nor the nanosecond. It is rhythm itself: rhythm in relationships. Intuitively, we know this and acknowledge it when we talk of biorhythms and someone falling into step with others or rejoining a conversation 'without missing a beat.' The foundations of human time, to this day, are not found in the tick-tocking of the clock but in the pulse of life flowing between people and within the body. Time is anchored in the beat of the heart and its systolic (compression) and diastolic (relaxation) phases.*
>
> —Heather Menzies, *No Time*

Rhythm, when measured, becomes tempo. Tempo is the rate of the recurring event. Time *tempo* differs from time *mode* in that tempo refers to the speed or pace at which someone within a culture accomplishes everyday tasks. Levine expands on Hall's research studying the relationship of various cultures by adding 'tempo'—looking at temporal idiosyncrasies of culture. "Adjusting to an alien tempo can pose as many difficulties as learning a foreign language," (p. 5). As Roger Newton writes in *Galileo's Pendulum* (2004), "Although biological mechanisms do not work with the accuracy or stability of modern clocks, a sense of time and its rhythms is built into the functioning of the human body. Our heart, with its beating pulse, is the clock-like internal rhythm we are most aware of" (p. 5).

Levine and his research teams measured the pace of life in thirty-one countries and thirty-six U.S. cities based on three variables which calculate the overall pace of life: walking speeds, postal times and clock accuracy. In the U.S. the postal time was changed to bank transaction speed and talking speed was added. Not surprisingly, he also discovered relationships between attitudes towards time and power/hierarchy. Those in power can, with impunity, keep those of lower status waiting.

Those in power have also tried, and often failed, to re-segment time, most notably, pre-empire Napoleon with his Revolutionary decimalized calendar and Stalin with his five-day week/six week month. In both cases, the standard Gregorian calendar was eventually reinstated because, "Temporal habits die hard… people will continue to battle unsolicited controls on their preferred, 'natural' pace of life" (Levine, 1997, p. 79).

Tempo patterns like walking speed, clock accuracy and transaction time intersect with, but do not necessarily overlap, cultural value pairs such as objective/relational and materialistic/quality of life, insofar as the manner in which the use or awareness of time affects any of these. In a similar fashion, a culture's time tempo may determine technology decisions regarding download speeds and completion status indicators. Time patterns, both mode and tempo, are intensely cultural. Gonzalez and Zimbardo note that, "every child learns a time perspective that is appropriate to the values and needs of his society" (in Guerrero, DeVito & Hecht, 1999, p. 227).

We must not over-extend our analysis of tempo, however. "There are dangers in making generalizations about the characteristics of places, particularly when they are directed at collective 'personalities' of their people," Levine cautions. "The notion of applying a single set of characteristics to an entire population, or to any group for that matter, makes for sloppy thinking" (p. xvii-xviii). That being said, Levine, like Hall, Hofstede and others, can't help but observe that people do exhibit time-focused stereotypes and that, "it would be naive to deny the existence of significant, overall differences between places and cultures" (p. xviii).

Time Mode and Tempo on the Web

In the process of reviewing these categorizations of time, mode and tempo, a certain pattern emerges. There appears to be a common thread between modern physics, and anthropology, a common dichotomy between Heisenberg's particle and wave, Levine's clock-time and event-time, and Hall's M-time and P-time.

In the early days of the Web, technological considerations resulted in monochronic designs largely because the technology had not yet progressed to the point where multiple activities could successfully co-exist in the same interface. Each task was often executed on its own webpage and the Home page became a table of contents that funneled users to the correct page for the desired action. The introduction of more sophisticated layouts, underlying software, databases and server query languages made it possible to offer the user multiple choices all in the same screen. Despite this increase in sophistication, monochronic websites continue to exist, including some of the most popular sites in the Web.

Google search is the classic example of a monochronic or linear design —there is only one thing you can do on its Webpage—search. Another example is Twitter where users, regardless of culture, have a limited set of actions: they can create short personal blog messages called 'tweets' and follow the tweets of others (Figure 6.1). The monochronic nature of Twitter limits the user to one function, the creation or consumption of one thing—tweets. Unlike Facebook, there are no games, no calendars, no separate channels of communication (commenting, messaging, status posting, poking, liking.)

Figure 6.1 Twitter's mono*chronic* linear interface

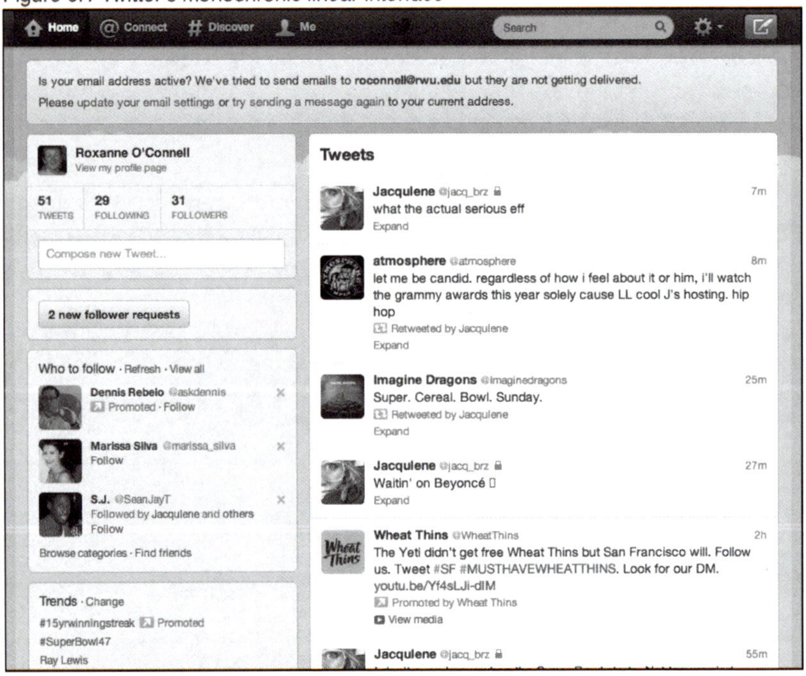

Polychronic designs are much more abundant on the Web. One example in particular may be an indicator that our monochronic Western U.S. culture is undergoing a shift: Facebook not only illustrates parallel navigation, it encourages polychronic behavior as users habitually juggle many activities at the same time and focus on relationship building and maintenance. This is a shift that bears watching.

Tempo is often exhibited in terms of tolerances: how tolerant users are of page download speeds or the time it takes for transactions to complete. The visual feedback users get from a website when something is processing may tell us something about the targeted culture's need for speed. While visual feedback, e.g., the [--- LOADING ---] message often seen on Flash™ webpages, is a standard usability consideration, some cultures deal more patiently with the ambiguity of poor status messaging than others. Fast tempo cultures will certainly abandon a site that appears to not be responding. Again, there may be more than cultural attitudes at work. Does the targeted user live in an area that is very advanced in available technology? Or does the user live in an area with slow network speed, frequent brown-outs and aging equipment? The more the owner and designer of the website know about the targeted user's location, the better the design will meet their cultural expectations.

Summary

Time mode and tempo expose two related but distinct aspects of a website's basic design structure. In this chapter I borrowed from Edward T. Hall's observations as presented in *The Dance of Life* and *Beyond Culture*, to look at time mode and a culture's orientation toward time. In *A Geography of Time*, Robert Levine expanded on Hall's research studying the relationship of various cultures by adding 'tempo'—looking at how fast people walk, how long transactions take at banks, post offices, train stations, and other temporal idiosyncrasies of culture. These relationships to culture are not always readily apparent in websites, but combined with examinations of other visual media, patterns can be perceived revealing subtleties that might attract or repel the target audience.

For those that want to explore other dimensions of time, I highly recommend Heather Menzies' *No Time: Stress and Crisis of Modern Life* and Mihaly Csikszentmihalyi's *Flow: The Psychology of Optimal Experience.* While these cultural time relationships, mode and tempo, are not always readily apparent in websites, and when they are, may be attributed to either cultural preferences, technology infrastructure or both, they might also be a characteristic or property of the way interaction is managed. Interaction is discussed more fully in Chapter 7.

Notes

1 Werner Heisenberg was a theoretical physicist known for developing two important theories in quantum physics: the Uncertainty Principle, or Heisenberg Principle, and the Observer Effect, which suggests that the act of observation will change the behavior of the phenomena being observed. While Heisenberg dealt with quantum mechanics, i.e., very small objects, and not human systems, they are analogous in that humans do behave and relate differently to phenomena on which they focus their attention.

2 As an interesting commentary on the immutability of ephemera on the Internet, the Google search "What is a facebook minute?" returned a definition from yourrelationshipedge.com, however the original post no longer exists.

Time Quadrants and Maps

M-time folks often like to use linear tools to map information. P-time folks find a mind map more conducive to non-linear, associative reflection. To experience the differences between M-time and P-time, map evidence you have found about your target culture's relationships to time using both a quadrant square and as a mind map. Do this twice, first using a fast tempo and then using a slow tempo. When you are done, write and reflect on how each exercise felt. You may want to establish which of these modes and tempos is your natural one before beginning.

Purpose: To experience a culture's Time *mode* and *tempo.*

Method 1 - Quadrant Square:

Quadrant squares help us see the relationship between two metrics. Joseph Luft and Harrington Ingham developed a quadrant square called the Johari window for developing awareness and understanding of a person's relationship to self and to others. Stephen Covey developed a quadrant for helping people understand and achieve their work-life balance. We can use a quadrant square to see if there are relationships between a culture's mode and tempo by mapping evidence we have found in the visual material we have collected, augmented by any written material we have found concerning our target culture.

	M-Time	P-Time
FAST		
SLOW		

Using one-word descriptors or keywords encourages brevity and specificity. If there is only one thing one can do on a given page, label it and put it in the P-Time column based on how much time it requires. For instance, you only have to click one button of two buttons, but you need to read several paragraphs to know if it's the correct button to select. You would label the action and put it in the M-Time/Slow quadrant.

Method 2 - Mind Map:

Mind mapping is a brainstorming/reflection tool refined by Tony Buzan in the 1970s. One starts by placing a 'focus' word or image in the center of a large piece of paper, and then branching out on 'radiants' in a non-hierarchical fashion that results in an image resembling a neuron or brain cell with all its synaptic connections. Using color and doodling are encouraged. In fact, this is what makes a mind map work—the subconscious synaptic associations we create between ideas, events, and information. There are several books and websites on mind mapping and I have, myself, written on Mind Mapping as a visual strategy to support critical thinking.

For this mind map, use paper and colored markers or pencils. Using a computer shifts your efforts to figuring out the software and away from a deep exploration of the topic.

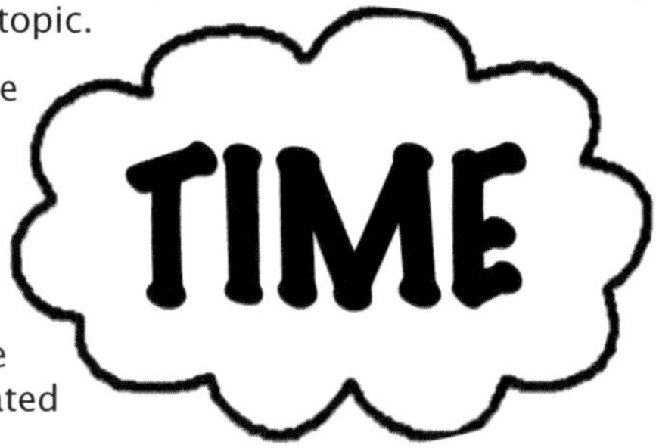

Place the word **TIME** in the middle as the focus. You *could* draw an image of a clock, but that might prejudice your thinking along monochronic, time-measuring lines. Use the radiants to brainstorm all of the observations you feel are related to the topic of TIME.

Keep to single words in ALL CAPS. Mixed case suggests hierarchy. Use the relative size of the whole word to suggest importance if you want emphasis. Don't try to categorize your ideas too narrowly. Just follow the natural, associative relationships that pop into your mind or onto the paper.

Doodle, add color and shading for emphasis. Let yourself get totally immersed.

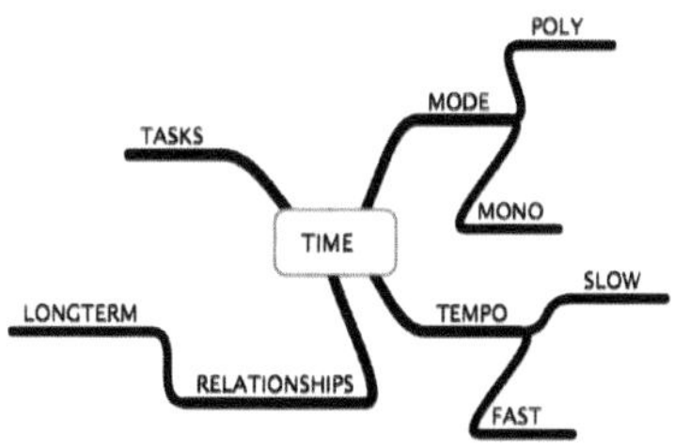

Do this for 20 minutes or so and then put the mind map aside for a couple of hours. Do something else unrelated to this exercise. Then come back and take a fresh look at what you have. Chances are you will have things to add—go ahead and do that.

Step back a bit and see if the mind map has any patterns to reveal.

When you have done both methods, perhaps in both slow and fast tempos, pause and reflect on how each felt. Experiencing and discovering your own personal mode and tempo will help you better understand and identify the temporal modes and rhythms of other cultures. Once you do that, you will be better situated to recognize what might be happening on a website (or in a film) that speaks to a culture's temporal preferences.

Chapter 7

Audience Interaction & Information Density

> *New media is interactive. In contrast to old media where the order of presentation is fixed, the user can now interact with a media object. In the process of interaction the user can choose which elements to display or which paths to follow, thus generating a unique work. In this way the user becomes the co-author of the work.*
>
> —Lev Manovich, *The Language of New Media*

In this chapter, we look at ways by which we can decide whether a culture prefers *active* or *passive* websites and the expectations concerning the flow of information. The terms active or passive are degrees of interactivity because all websites are interactive. An extremely active website has myriad ways and means for the user: links to follow, polls to take, forms to fill, apps to download, all from the home page and perhaps many other pages on the site as well. Facebook is a highly active website with many prompts for things to do, nearly all of them targeted at either building content on your Facebook page, growing your network of 'friends,' or getting your contact information and those of your 'friends' to advertisers through clever ads and apps (Figure 7.1). Facebook would be at one extreme of the active/passive continuum. The other extreme of interaction would be Google's search home page where there is only one thing you can do—search (Figure 7.2).

Western, and in particular American, corporate websites tend to be quite active, using lots of Flash™ media with images that pop up at the viewer when the cursor passes over that part of the screen. However, not all cultures care for busy, active websites. This is most evident when we compare the U.S. version of a multinational site to its localized versions for other countries and cultures. There, we often see a decrease in interactivity, and we can speculate as to whether this is a technical

Figure 7.1 Facebook as an example of very active interaction

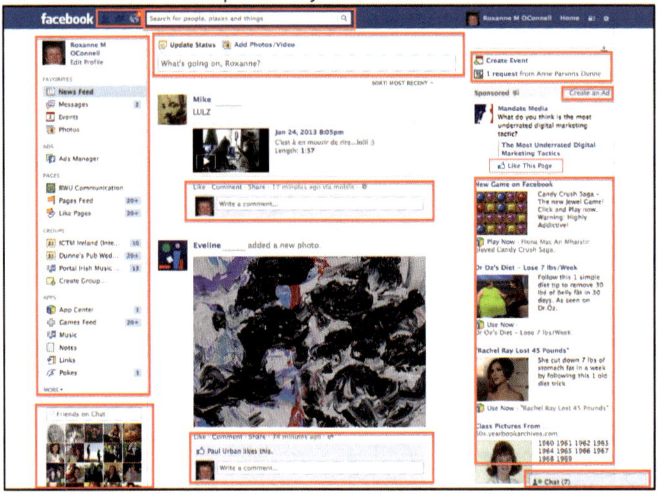

Figure 7.2 Google as an example of passive interaction emphasized by the amount of negative (white) space.

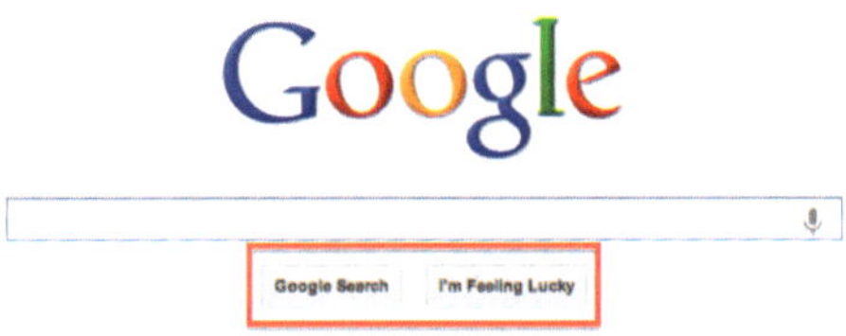

consideration or an aesthetic one. Either way, it provides important information for the researcher and designer.

Information flow is related to interaction in that once an interaction scheme has been designed, the next design decision is what will the user see when they have clicked? What kinds of channels and gates must be set up so that the user's subconscious expectations around access and privacy are met, as well as the site owner's desire for information control. It helps enormously to know the purpose of the website, the kinds of information it makes available and the kinds of information it asks for in return.

To go back to our two examples, Facebook™ presents itself as a personal space for you and your friends. Depending on your personal sharing settings, you can open or limit the amount of information that is shared with the world through your Facebook page. Facebook's underlying purpose, however, is to collect as much data as it can for its advertisers and corporate-driven changes to user privacy settings have often generated much criticism. The many confrontations the Facebook™ organization has had with its users is testament to the fact that they continuously push the boundaries regarding privacy and access. Google, on the other hand, has no user login or barrier to using its search engine. They express this attitude toward information in their mission statement: "Google's mission is to organize the world's information and make it universally accessible and useful" (http://www.google.com/about/company). Some cultures, or to be more precise, some governments like China, have had difficulty with this idea that information should be universally accessible and the struggle to control the flow of information in those parts of the world continues.

Audience Interaction

Whenever possible we try to assess if a culture prefers interactive or passive websites. Interaction design is an essential field in the creation of all sorts of artifacts. Gillian Crampton Smith, in the foreword to Moggridge's *Designing Interactions* (2007), writes:

> To designing for usability, utility, satisfaction, and communicative qualities, we should add a fifth imperative: designing for *sociability*. When IT systems fail to support the social aspect of work and leisure, when they dehumanize and de-civilize our relationship with each other, they impoverish the rich social web in which we live and operate. (p. xiv)

Designing for a cultural aesthetic spans the concerns of satisfaction, communication and social interaction.

Even within a particular *geo-ethnic* culture there may be sub-cultures that have more in common with their *functional* counterparts based on the kind of website and activity in which they are engaged. A good example would be the online gaming culture and its close cousin, the alternative reality culture. Users of these kinds of websites will have radically different interaction expectations based on the website's purpose. While these sites might appear to the uninitiated as similar because both deal with fantasized worlds, they are not approached or used the same way and the cultural norms you might observe in virtual community sites like SecondLife™ or China's HiPiHi™, where users create their ideal living environment, will be vastly different to those you will experience in World of Warcraft™, a role playing game of adventures and conquest. There are distinct cultural norms in these imagined communities. Several studies have been done on various cultural behaviors and these virtual worlds have become, not only subjects of study, but also milieus in which to conduct ethnographic research (see Kim Gregson's chapter "Teaching Qualitative Research Methods in SecondLife," O'Connell, 2012).

Interaction is the heart of the Web. The ability to click on a link or button is what makes the Web function the way it does. Lev Manovich (2001) reminds us that "Old media involved a human creator who manually assembled textual, visual, and/or audio elements into a particular composition or sequence. This sequence was stored in some material, its order determined once and for all" (p. 36). He maintains that a new media object, like a website and its pages, are not "fixed once and for all" but, rather, can exist in multiple versions as the user experiences them and he likens them to postindustrial artifacts that rely on "just in time" or "production on demand" delivery models.

There are three interaction types that Manovich describes in his discussion of Web architecture and the variability of media behavior it makes possible. At one extreme we have simple *serial* linking. Think of a slide show embedded in a website. You move forward or backward along a serial chain of images. You cannot skip through to an image halfway through the stack, although you can perhaps click through rapidly. The order of the images is fixed, not unlike 'old' media. Some cultures may prefer this level of predictability and order.

Another interaction type is *branching* or *menu-based* interactivity. This allows a user to move through a network of related material. According to Manovich, "When the user reaches a particular object, the program [or website] presents her with choices and allows her to choose among them" (2001, p. 38). A good example would be a shopping website where a user is given the choice between "Continue Shopping," "View Cart," or "Pay". Choosing "Continue Shopping" brings the user back up a branch of the site 'tree'. Branching is a typical interaction strategy for information seeking. On a news site, a user can click on the linked headline or image of a story to view the full story. They might then click through several pages of the story and then use the top level menu to return to the front page. It is not a random interaction but one that follows a particular path, although the user always has the option to jump to another branch.

The third interaction type is *hypermedia*, what used to be called hypertext when the Web was largely word driven. By following hypermedia links a user can be brought anywhere on the Web, whether the link had a connection to the narrative flow of the original website or not. In 2001, hyperlinks were cognitively obvious. "A hyperlink creates a connection between two elements for example, between two words in two different pages or a sentence on one page and an image in another, or two different places within the same page" (p. 41). But in a hypermediated world with billions of webpages distributed over an estimated 644 million websites in the world, according to the San Francisco Chronicle, it doesn't take long to click your way far from where you thought you were going. For some people this can create a sensation of information overload and "How did I get here?" disorientation. Hours can go by leading to new conceptions of time such as a Facebook minute.

Manovich cautions us to think of interactivity, not just in its transactional sense of clicking on a button or link, but also as it impacts the psychological processes of hypothesis formation, recall, anticipation and identification. Just as the avant-garde movements in art and cinema used montage and other forms of abstraction to force the viewer to make the leap and bridge the gap in the narrative, navigating hyperlinked media appears to give the user control over his or her choices. In reality, "we are asked to follow pre-programmed, objectively existing associations" (p.61). What the researcher of cultural aesthetics must take into consideration when evaluating a culture's interaction preferences is to what extent are the preferred modes of interaction shaped by the designer and which are shaped by the culture of the viewer.

Interaction is relevant to several other pattern concepts. Most of these concepts deal with control: control over where we are going, over what we expect to see or will be allowed to see, and control over how much information we need in order to feel comfortable with what we are doing, especially when that interaction deals with commerce or the exchange of personal information.

Perhaps the most interesting technology crossroad in audience interaction is where new haptic interfaces intersect with our more traditional attitudes toward information flow. The introduction of the iPad and mobile media and computing devices, like our smartphones, have changed the landscape of interaction. The highly intuitive pinching and swiping of touch-screen enabled interfaces are extending our notion of 'user' to those much younger in age. YouTube videos of toddlers interacting with traditional media, like magazines, as if they were iPads are amusing and yet, at the same time, somewhat disquieting (CBS, 2011). Very young children have not yet been acculturated with regard to power and the flow of information. If any child can quickly and intuitively grasp how to get what it wants in the digital media sphere using these new interfaces, will a culture's interaction preferences as they relate to controlling the flow of information evaporate? For researchers in material culture, the coming few years will be a very interesting period of time.

Information Density

Information density is the degree to which a website complies with a culture's accepted power distance, a cultural dimension defined by

Hofstede as "the extent to which the less powerful members of institutions and organizations within a country expect and accept that power is distributed unequally" (1997, p. 28). It is also related to E. T. Hall's concept of information flow (1987). The difference is that flow concerns the speed and structure of message code—a headline is fast, the whole article is slow—whereas, power distance is a question of access to information and who is authorized to access it. Does one need to be a member, subscriber or employee to get at that information? Or is it readily available to any user from any location?

At first blush, this may appear to be one-sided: the powerful, the ones in control, the boardroom elite against the common person, the user. However, in today's social media society, who gets access to what information is much more complex and not at all transparent. The fact is that it is the job of every commercial website to get as much personal information about you and your friends out of you as they can in the few minutes you spend on their site. To this end, they offer all kinds of interactive elements like polls and FREE memberships to keep you on the site and clicking. The data collected about you and the click-stream you created to innocently share that information with them—the site you came from, what you clicked on while you were there and where you went once you left—are all bits of data that can be sold to a data aggregator who can then repackage that information and sell it again to marketers looking for what will make you look and buy.

Douglas Rushkoff has been documenting our relationship with the advertising/marketing world since 2001 with "Merchants of Cool" and "The Persuaders"—a truly scary documentary about our vulnerability to marketing manipulation. He wrote regarding a CNN report on the National Security Agency's surveillance programs:

> Your life and pursuits are less important than the statistical profile of the way you use your digital devices. This is the world of big data.
>
> I remember the days when talking about such possibilities was considered conspiracy theory or paranoia. Many of us imagined a future in which people would be planted with chips that monitor our conversations and whereabouts. Perhaps we'd even accept such tagging voluntarily, if it meant being able to track down our children in the unlikely event of a kidnapping. But such extraordinary measures proved unnecessary; we're all walking around with tracking devices in our pockets, which are capable not simply of broadcasting our phone calls but our physical locations,

> our movements, our interests - and then to tie all this data to our consumer profiles, credit histories… everything. (rushkoff.com, n.p.)

Unfortunately, far too many of us pay little attention to all of those Privacy Policies and Terms and Conditions websites are required to show us. More often than not we click and accept them without reading them.

Nevertheless, the power on the Web, if one chooses to use it, is with the user most of the time. The question for the researcher is whether one can tell if the target culture has a high or low power distance. This is where knowledge of the language used on the website is most useful. Power distance will be evident in the 'voice' used by the website when addressing the user. As a way to experience that, go to your national government's website, to the law or tax divisions, or to your local registry of motor vehicles. 'Listen' to the tone. Most government entities, but particularly those involved in law enforcement, regulation or tax collection, try to establish a high power distance, even in low power distance cultures. Even in the most egalitarian of cultures there is hierarchy.

Any website that requires a form will also exhibit a high power distance. The more sensitive or private the information being requested, the higher the distance in terms of tone and voice. Websites that are situated in low power distance cultures tend to bury those pages that deal with privacy, policies and terms of use. People in low power distance cultures react negatively to suggestions of inequality in even online business or information relationships.

Information density, on the other hand, is not about how formal the rhetoric on a website is, but how much or how little information is available to the user and under what circumstances. In news websites that have a long tradition for investigative journalism, not only will the news stories be more lengthy than in the sites of their tabloid cousins, but there will also be links to related articles, many of equal length and depth. In information sites like Wikipedia, the community that has helped build it and maintain it is committed to creating as incredibly rich, dense and complete a body of information as can be found anywhere in both the virtual and dirt worlds. NASA.gov is another example of a site with dense information that is not just verbal but also visual and dynamic.

In high power distance cultures, the information density in the open access areas of a website will be low, unless there is a compelling regulatory reason for supplying more information. In which case, the information will be voiced in a more formal, authoritarian tone. In low power distance cultures, the information density in open access areas will be high, unless there is a compelling commercial reason for putting up a gate. In the magazine marketing world there are two ways to monetize the reader: find out something about them you can sell to a list or data aggregator (called controlled circulation) or have the user pay a fee to access the information (called paid subscription). The same model applies to websites. Facebook™ and other social media sites do not charge you for being a member, but they collect every piece of data they possibly can once you have logged on. All of this is then fed to advertisers and their ads inevitably, and somewhat creepily, appear on your Facebook™ page. Other sites give you the option—have the page be cluttered up by indiscriminate ads or upgrade to the Pro version of the site.

Summary

Audience participation is what makes the Web the engaging, informative, entertaining and challenging force that it is in our lives. New modes of interaction challenge us to relearn how to navigate the Web. New attitudes toward information and its availability and dissemination have reshaped how we look at information and media content. As an area for further research, the manner in which these changes in audience interaction and information flow are influenced by our many cultures, and how our cultures are, in turn, influenced by these technologies and what they make available to us, is rich with possibilities. At the moment, these technologies are largely visual, and so we rely on what our eyes tell us about how websites behave (interaction) and what they reveal to us (information flow). In the not-too-distant future, we will also be able to use our other senses more fully, but for now, our eyes—and increasingly our ears—provide us with the clues to how a culture wants to participate on the Web.

But audience participation is controlled and steered by information flow and power relationships. Commercial websites try to manage access and participation in order to monetize their sites, either through free memberships or through paid subscriptions to the upgraded version of the site. Although perhaps not universally true, the law enforcement, regulatory and tax divisions of government sites generally exhibit a high power distance tone as does any form that requires sensitive or very private information, especially of a financial nature. As friendly as a low power distance culture might want that language to be, the liabilities associated with language less legal and formal tend to prohibit more user friendly tones.

Affordance Testing

Buttons and links do different things. Buttons, like INSTALL, SEARCH, or BUY NOW, execute a function. They begin a chain of events usually governed by the underlying applications on the website. Links, on the other hand, are simply navigational. They bring you somewhere. Both links and buttons have visual cues that telegraph to the user what should happen when either of them is clicked. Good design exploits these visual cues so that the user isn't surprised by what happens when he or she clicks. Bad design is usually the result of not using the button/link conventions appropriately.

This simple exercise helps you catalog the interactive mechanisms of your target culture's websites.

Purpose: To better understand a website's interaction and flow

Method: Print, in grayscale, a website's home page and several second level pages. Grayscale removes any color cues and leaves you with only those cues that use shape, size, shading, dimensionality and emphasis (bolding and underlining).

- Using a yellow highlighter, highlight everything you think is something you can click on to execute a function (button-like action).
- Using a different colored highlighter, highlight everything you think is a navigational link.
- Count how many things one can execute on the page (interaction).
- Count how many clicks it takes navigationally to get to key information (flow).

When you have completed your highlighting, go to the website and click on each of the respective interactive and navigational elements.

This exercise can also be used to determine a website's usability when you are looking at *uncertainty avoidance*. Clicks that bring a user to places they didn't expect or which launch interactions they were unprepared for trigger feelings of unease in people with high uncertainty avoidance preferences.

*This exercise is a common usability test. A more extensive explanation of affordance and the method for testing it can be found in the "Website Usability" chapter in *Visualizing the Web* (O'Connell, 2010, pp. 132–134).

Chapter 8

Cultural Attitudes & Values

> *A people's ethos is the tone, character, and quality of their life, its moral and aesthetic style and mood; it is the underlying attitude toward themselves and their world that life reflects. Their world view is their picture of the way things in sheer actuality are, their concept of nature, of self, of society.*
>
> —Clifford Geertz, *The Interpretations of Cultures*

In his seminal work, *Beliefs, Attitudes and Values* (1968) Rokeach defines an attitude as "a relatively enduring organization of beliefs around an object or situation predisposing one to respond in some preferential manner" (p. 112). A culture's values, and the visual cues that reveal them, are a culture's "learned organization of rules for making choices and for resolving conflicts" (p. 161). The cultural values and attitudes we discuss as patterns in this chapter, as with all of the cultural patterns we have been exploring, come with a caveat. Samovar, Porter and Stefani express it best in *Communication Between Cultures* (1998): "The value of the culture may not be the value of all individuals within the culture.... We are all much more than our culture." However, they do maintain that "culture has the strongest influence on our communication behavior" (p. 61) and that presumably extends to our communication behavior on the Web.

While many of the patterns we have already discussed can be evidenced in the visual and interactive elements of a website, cultural values and attitudes are more nuanced. Is the image we are looking at evidence of a high- or low-context culture? Is the context inflected by collectivist or individualist attitudes? Are the relationships depicted in the image appealing to a relational orientation or an emotional affect or both? Each of these descriptors acts as color and shading in our pattern language. Therefore, the best approach to each of these attitude and value pairs is to see each pair, not as polar opposites, but as positions that interact with other attitude and value pairs and that vary in intensity as a result of those interactions.

We must also refrain from essentializing the entire population of any country or region by attributing any of these patterns to all who live there. Far more interesting is the possibility that outliers in a large group of websites being examined might actually be targeting a cultural population that is not the dominant culture of that place. Using the Grounded Theory approach of keeping open to new and emerging patterns better serves both the researcher and the designer. As Samovar et al point out, "there are literally hundreds of cultural patterns found in every culture" (p. 61). We are trying to identify those patterns that speak to a specific target audience. Before assuming that a website is missing the target of the greater dominant culture, we can raise the question of whether there is an embedded or fringe audience that is possibly the target. If we see that there are indeed links into the site, traffic being generated and interactivity evidenced, we must assume that the needs of a group of users are being met. Who they are is for the researcher and designer to discover.

Finally, as the project that led to this work made very clear, technologies change, cultures change, and therefore it is not outside the realm of possibility that a culture's values can shift and change. Modern Japan is an excellent example of aesthetic pattern change. I had grown comfortable over the years using 'Scandinavian Design' and 'Japanese Kanso' (the removal of clutter) as examples of different approaches to white space—that is, until my students kept bringing me modern Japanese websites that were very cluttered and dynamic. We observed a shift in patterns from spare, poetic, zen-like functionality to the colorful, crowded, busy, dynamic visual cacophony that is often now seen in sites targeted at younger audiences. The cultural aesthetic had not actually shifted from one to the other; it had split into two opposing patterns. Both coexist, just as the Shinto shrine coexists with downtown Tokyo, but the newer pattern is gaining dominance.

In this chapter we look at a collection of value pairs and cultural attitudes that provide further insights into a culture's aesthetic. Each value pair is a continuum, however a given culture's dominant value will situate itself on one side or the other of the mid-point. The first value pair deals with social or cultural context, or how the members of that culture code and decode messages. E. T. Hall used the term high-context or low-context to describe which form of communication, or coding/decoding, is valued by a culture. Geert Hofstede, through his work researching cultural values in the work place in the 1980s, gives us several terms to describe cultural behaviors, attitudes and values. Power distance has already been mentioned in Chapter 7 where we discuss Information Flow. Uncertainty Avoidance is a behavior we discuss more fully in Chapter 9 as it may be a driver in a culture's emphasis on usability. The remaining value pair he characterized as masculinity/femininity. Not to be confused with gender, Hofstede's masculine-feminine value trait has often been criticized for lacking enough granularity or nuance to truly capture these behavior complexities. I have broken this one pair into three interrelated value orientation pairs that give us more flexibility for identifying patterns: rational/emotional, objective/relational, and materialist/quality of life. Finally, Hofstede articulates a cultural attitude, collectivist or individualist, that is often discovered in a culture's use of imagery or photographic subject. He suggests that a culture's dominant attitude can be characterized as either collectivist or individualist depending on where the culture's intrinsic and extrinsic motivations originate: out of concern for the welfare of the group or for that of the individual.

Closely related to the *content* discussion in Chapter 5, examining cultural cues concerning imagery (what is in the images), context (their meaning), and attitude (how they are expressed) adds greatly to the examination of a culture's visual aesthetic. As with the chapter on Time mode and tempo, this chapter examines highly nuanced visual cues that deeply resonate with the intended audience. Much as using local idioms in a verbal language bridges the formality gap, understanding these visual cultural cues helps the audience 'know' that this is for them. For instance, a culture's lifestyle value (materialist or quality of life) is a subtle, but critical, cue often ignored in large multinational websites where brand images often dominate. When the audience is part of that brand's dominant culture, this is less important. However, in cultures that may not share the same underlying values, the visual cues must match the cultural ones to be successful in conveying the brand's message.

Examples in this section are, by necessity, anecdotal and are meant to illustrate the point rather than provide hard evidence. Cultural values are, overall, too nuanced and interrelated for highly specific categorization, although many scholars have spent entire careers in an attempt to create ways to measure them. The value pairs and cultural attitudes we discuss overlap, sometimes significantly, and the researcher is better served using these pattern concepts as terms that describe lenses or perspectives rather than hard divisions.

Cultural Context: High or Low

Hall's concept of context describes how people in a culture perceive and communicate meaning. This pattern can be connected to the visual/verbal modes of content as context is determined by which mode is being used to communicate: either through the settings in which communication happens or through the actual words being used. High context cultures require a high degree of homogeneity in traditions, rituals and practices. Even low-context cultures have high-context pockets. To illustrate this point to students, I often use the example of flagging down a bus at a city bus stop in Dublin, Ireland. It took me ages to figure out that, even though I was standing at the correct bus stop, I had to put my hand out so that the driver would know I wanted that particular bus. I had not accounted for the fact that many different routes might use the same bus stop and the only way a driver would know I wanted to get on was if I flagged him down. Who finally clued me in? A six-year old, much to my embarrassment. Apparently, Dublin children learn at a very early age how to get a bus to stop, city buses being a major mode of transportation for the residents. Various sociologists have listed Ireland as a low context culture, but as an American who lived there for several years, I would absolutely categorize Ireland as high context. The Irish have a strong sense of tradition and history, have close-knit extended families, and knowledge is situational and relational. For instance, it is considered impolite to respond in the affirmative the first time or two one is asked if they would like a cup of tea or coffee. One should refuse until pressed by the host the third time, usually with the following response, "Ah, well, if you're making it for yourself, I wouldn't mind a cup." No one tells you this, but were you to accept the first offer, however kindly, you could be considered a rude and uncouth guest. Perhaps the most telling indication about the high context of Ireland's culture is that whenever I complained about not being told something, I'd hear, "But, sure, everyone *knows* that!"

In websites, patterns can be evaluated as high or low context by looking at how much of the message is coded in explicit words and how much is 'contextual' to the location and type of imagery, navigation, and functionality. When words are used to instruct or navigate, the message is low-context. When the non-verbal elements on the page are presumed to carry the message of what to do and where to go, the message can be categorized as high-context.

Attitude: Collectivist or Individualist

Individualist cultures are driven by internal motivations. Collectivist cultures are driven by the expectations of others, particularly those of community and those in authority. This attitude pair is often combined with other value pairs. For instance, socially collectivist cultures may also be polychronic and/or high context. Collectivist cultures fall into a relational orientation that is based either on family or work/community groupings. Cultural attitudes and values are nothing if not complex.

The relationships between people in the images displayed in Websites are potential indicators of that culture's attitudes and values. Often in

East Asian cultures where the work or community groupings are placed above individuals or family, people in images will be anonymous or too distant in the image to be easily identified, unless they are political figures or national celebrities. Images of families, particularly multigenerational groupings, resonate strongly with Latino cultures. Some years ago, I was part of a 'globalization' team to help redesign the website for an entrepreneurial company that specialized in providing telephone cards for immigrants in the U.S. Latino market. The website was largely 'techno' and noticeably devoid of images of people. The color scheme was a very 'corporate' blue and steel gray on white. It lacked warmth and vibrancy and the copy emphasized low-cost instead of playing to the product's strength—its ability to allow people with low or no credit to have prepaid phone cards that helped them keep in touch with their family in Mexico, Guatemala or Brazil. Our re-worked design used warm, vibrant 'call to action' colors like deep brown reds and ochre yellows, and lots of images of families enjoying celebrations and each other's company. We helped craft copy that emphasized a 'cultural' value proposition of connectedness, security and ease of use.

In summary, collectivist versus individualist cultural attitudes are evidenced in the kinds of images of people we see. Collectivist cultures often use images of large groups of people where the individual is not necessarily foregrounded. This contrasts with the Western concept of rugged individualism, which, when brought to its extreme, is evidenced through images of the media celebrity as much as the political candidate or successful entrepreneur. In an individualist society, the entrepreneur achieved success on his or her own. In a collectivist society success is conceived and perceived as a group effort.

Affect: Rational or Emotional

The rational/emotional, or *logos/pathos*, value pair is a subset of Hofstede's masculinity/femininity cultural attitude pair. Hofstede based his analysis on behaviors that fell into one of these polar opposites: assertiveness (masculine) versus modesty (feminine). Disengaging the notion of gender roles from the terms masculine and feminine is awkward at best. This is further complicated when one examines the population he was studying. Hofstede was working for IBM during the 1970s and 80s—at that time, the largest multinational computer company in the world. Management and executive tiers of the company were dominated by men and, presumably, response rates to his IBM questionnaire would have been similarly gender biased. Rather than get entrapped in whether or not cultures can be masculine or feminine, we can more profitably look at non-gender situated value pairs that better articulate what we are seeing when we look at a culture's aesthetics on the Web.

People and cultures that show a preference for facts or fact-based arguments can be characterized as valuing the rational. Those that show a preference for drama, exuberance, or emotional appeals can be characterized as valuing the emotional. Imagery that is used as information or decoration may provide the clues to which side of the continuum a culture is situated. In addition, rational cultures will likely rely more heavily on verbal content, perceiving it to be more direct and concrete and less open to interpretation. Images are, by their nature, more relative in interpretation. Roland Barthes (1957) reminds us that semiotic or photographic truth is influenced by culture and inflected by a culture's myths, values and ideologies. There is truth in the image but it is rarely the same truth for all.

Orientation: Objective or Relational

While some may find the objective/relational pair similar to rational/emotional, they are quite different in subtle ways. While rational/emotional is a resonating attitude or approach, the objective/relational pair is an orientation toward decision making. Websites that emphasize data driven reasons for making a purchase, such as competitive features or price point, are orienting the decision toward objective data. Consumer comparison websites are a good example of a site utilizing an objective orientation. Websites that rely on viral marketing are orienting the purchase decision toward the buyer's relationships: "Look what your friend just bought!" Facebook, Pinterest and Twitter are excellent examples of this new relational approach to marketing.

Hofstede's concept of collectivist versus individualistic cultures is also reflected in the objective/relational value pair. Collectivist cultures are societies "in which people from birth onwards are integrated into strong, cohesive ingroups" (1997, p. 260) Individualist cultures are societies where "ties between individuals are loose" and where one relies on oneself or one's immediate family (p. 261). As large multinational social media sites begin to dominate the Websphere, we may see a shift in how collectivist/individualist and objective/relational pairs manifest themselves within the cultures where these social media sites are widely integrated and adopted.

Lifestyle: Materialist or Quality of Life

In addition to the cultural attitudes and values outlined above, we can also look at the imagery displayed on a website in terms of the lifestyle that is depicted. Is the imagery based on *materialism* or *quality of life*? While an emotional attitude and a relational orientation is often coupled with a desire for quality of life in what Hofstede would have termed the female gender role, emotion is also used in conjunction with materialist values, most often in marketing luxury items, where quality of life is measured in materialistic terms. However, websites that are targeted at cultures, such as the Scandinavian countries, where people traditionally have a long view and are conservative regarding resource utilization, portray a different kind of 'quality of life,' one that favors family and friends over status and power. Imagery and design on these websites convey a 'just enough' lifestyle.

Summary

The cultural attitudes and values that have long been the subject of research by social psychologists and anthropologists are important clues to a culture's visual aesthetic. These attitudes and values are found most often in the kind of imagery that is prevalent on a website. They are also indicated by a subtle preference for words over pictures or the reverse. Matching what we see to these cultural value and attitude pairs helps us better see the depth and complexity of a target culture. We are better able to identify where the resonances lie and why one type of imagery speaks more powerfully to the target audience than does another. The extensive work that has been done by scholars such as E. T. Hall, Geert Hofstede and Milton Rokeach provide additional data against which to check our observations. However, we are wise to keep in mind that much of this research predates the Web itself and does not take into account any cultural hybridization or shifts in social practice that may have occurred as a result of emerging and popular social media. In addition, the populations measured in those studies may not be a good match for the target audiences we are trying to study or design for. Our job is to compare and contrast what we are seeing, allow patterns to emerge, and then use the best match of terms in our pattern language to describe what we are seeing.

A Day in the Life of My Cultural Twin

Purpose: To delve deeper into the cultural meaning of what we see.

What we see or read has two meanings: the denotative, which is its literal meaning, and the connotative, which is the culturally situated meaning. The semiotician, philosopher and critic Roland Barthes used the term *myth* to refer to the cultural values, attitudes and beliefs that are expressed through connotation. Sturken and Cartwright (2009) explain, "For Barthes, myth is the hidden set of rules and conventions through which meanings, which are specific to certain groups, are made to seem universal and given for a whole society. Myth thus allows the connotative meaning of a particular thing or image to appear to be denotative, literal, or natural" (p. 20).

- How will we know when we are looking at a denotative element or a connotative element?
- Which beliefs, attitudes or values govern the perceived connotation?
- Do any beliefs, attitudes or values speak for the whole culture?
- What *myths* do we have about this culture?
- What *myths* does this culture communicate about itself?

Method: To explore these questions do the following directed writing assignment and find images from your websites and ads to illustrate it:

- Write three to five pages describing, in detail, a day in the life of your cultural twin. Your twin can be of the same or different sex from yourself.
- Write down from the moment your twin awakes everything that she/he sees and does, the people she/he meets, the place in which she/he lives, works, plays, studies, rests, socializes.
- Continue up to the moment your twin closes his/her eyes and goes to sleep.
- Pay attention to the patterns of the day like eating, personal hygiene, exercise (or not). What products are used? What are the colors of the walls and furnishings? What is on the bedside table—if there is one? What are the small decisions your twin makes throughout the course of the day?
- Put the assignment aside for 24 hours.

- Re-read the assignment highlighting those descriptions you may have had prior to your visual explorations of your target culture (*myths you had about this culture*.)
- In a different color, highlight those descriptions or details you collected from your research (*myths this culture communicates about itself*.)

*This exercise is adapted from a writing assignment created in the 1990s by Mercia Tapping called, "A Day in the Life of My Successful Twin." The original purpose of the assignment was to help people who had recently been made redundant (laid off) imagine what their lives could be, unencumbered by any perceived limitations. I often use this exercise with students who are trying to imagine what they will do after graduation. I have also used the version outlined above in an ethnography course to help students 'put on the skin' of their subject and use 'thick' description.

Chapter 9

Usability, Uncertainty & Ambiguity

> *On the Web, usability is a necessary condition for survival. If a website is difficult to use, people leave. If the home page fails to clearly state what a company offers and what users can do on the site, people leave. If users get lost on a website, they leave. If a website's information is hard to read or doesn't answer users' key questions, they leave. Note a pattern here? There's no such thing as a user reading a website manual or otherwise spending much time trying to figure out an interface. There are plenty of other websites available; leaving is the first line of defense when users encounter a difficulty.*
>
> —Jakob Nielsen

According to noted usability guru Jakob Nielsen, *usability* is defined as "a quality attribute that assesses how easy user interfaces are to use" (2012, n.p.). Nielsen outlines five quality components that define just how usable an interface is, including: learnability (is it easy to learn?); efficiency (how quickly can someone perform a task once they have learned the interface?); memorability (is it easy to recall how to use the interface after gaps in time?); errors (how many and how severe are the errors users typically make with the interface and how easily can they recover?); and satisfaction (how aesthetically pleasant is the user's experience with the interface?). He also mentions a key component *utility*, or does it do what the users need? For the purposes of this chapter, we will put more emphasis on the notion of errors and their cultural interpretation.

Usability

First, let me state without reservation: every website, every application—indeed everything we put our eyes on and hands to—should be *usable.* Anything that doesn't do what the user expects it will do based on the visual and haptic clues provided is the result of bad design. Period.

Unless, of course, the purpose of the design is to puzzle, frustrate or, in some way challenge the norms of the user as in brain teasers or games. In these cases, the challenge is to make sure that the obstacles that face the user are by *design* and not by carelessness or shoddy workmanship.

While usability is a critical consideration for any technology, platform or medium, in some cultures the possibility of error carries more weight. Hofstede's concept of *uncertainty avoidance* is useful here: a desire to avoid the discomfort of ambiguity or uncertainty or the possibility of making errors. Nowhere do we encounter the possibility of making errors as often as when we are dealing with computers, software and the Web. This chapter outlines how to evaluate the usability of a website in order to assess the target culture's tolerance for ambiguity. A culture that values 'saving face' will have a low tolerance for errors or ambiguity while another culture, for instance the gaming culture, may have a very high tolerance for errors as these contribute to learning and winning the game. Therefore, when there are many options available to the user, they will naturally go to those sites that give them the highest level of confidence in their usability based on the user's tolerance for ambiguity and uncertainty.

Uncertainty

How is uncertainty avoidance, or the discomfort with ambiguity, measured within a culture? Hofstede's questionnaire focused on three unrelated questions that—at the country/culture level—pointed to a behavioral anxiety. Furthermore, that behavioral anxiety, while perhaps not expressed the same way from person to person, remained consistent within a particular culture. The first question concerned how much stress the respondent felt at work. The second question concerned 'rule orientation'—the belief that company rules should not be broken, "even when the employee thinks it is in the best interest of the company" (1997, p. 112). The third question asked how long the respondent felt they would remain working at IBM, the corporation from which Hofstede drew his research participants. Hofstede initially found it puzzling that someone who experienced high levels of stress at work would also express a desire to work at that company until they retired. This is a case where looking at the data *at the aggregate level* helps tell a more complete story. He writes, "Such reactions need not be found within the same *persons,* but only statistically more often in the same *society*" (p. 112). To confuse the individual level of stress or anxiety with that expressed as a pattern in a society constitutes an ecological fallacy.

Uncertainty avoidance is also related to 'saving face,' a sociologic concept that E. T. Hall connects to the concept of context. High context cultures, which focus on in-group behaviors, tend to be more invested in face saving than are low context cultures such as the United States and Australia, cultures that are also individualist in attitude. Other sociologists that have explored the notion of 'face'—saving, losing, presenting—include Erving Goffman (1955, 1959, 1967) and Bert Brown (1977). Websites that are insufficiently usable so as to allow users to make needless mistakes, or worse, make them unable to complete their tasks, increase uncertainty and make people feel stupid, or lose face. For people in cultures where ambiguity or uncertainty causes high levels of stress or where losing face is not tolerated, websites that were needlessly unusable would not find much favor. In a Darwinian sense, only intuitive, responsive websites with a great deal of helpful documentation and support would attract and keep users in these cultures.

Measuring Usability

How do we determine if a site is usable, tolerably ambiguous or just badly designed? Usability, or in this case, error avoidance, can be evaluated by looking at the functionality and predictability of navigation, the clarity and functionality of online forms, order processing and online information collection, and lastly the presence, quantity, quality of and ease of use for online help and site support. This last item, help and support, might not get much activity in a highly intuitive and usable site, so its presence as part of the site may be an indicator of the designer addressing the cultural needs of a target audience with high levels of uncertainty avoidance.

Singh and Pereira (2005) catalog a list of features a usable website should have that include:

- customer service elements such as Frequently Asked Question sections (FAQs), online help, and customer contact information;
- guided navigation schemes that provide site maps and where links are overt, navigation is consistent and accommodates movement backward, forward, up and down;
- a tradition-based theme, emphasizing reliability, history and multi-generational relationships;
- local store information for those who wish to research on the web but buy from local merchants;
- local terminology, emphasizing 'in-group' communication;
- free trials and downloads;
- toll-free phone numbers;

- and, lastly, testimonials from customers and transaction security indicators like seals of approval, certifications and appropriate security checkpoints in any process involving the exchange of money or goods.

Both the *Boston Globe* and *Boston Globe Media Services*, the former, the original newspaper company, and the latter, the media company into which it has grown, have many, if not all, of these features on both of their sites—each geared to their respective target audiences (Figure 9.1).

Figure 9.1 The "About Us" page for *Boston Globe Media Services*. The boxed areas are examples of Singh and Pereira's usable website features.

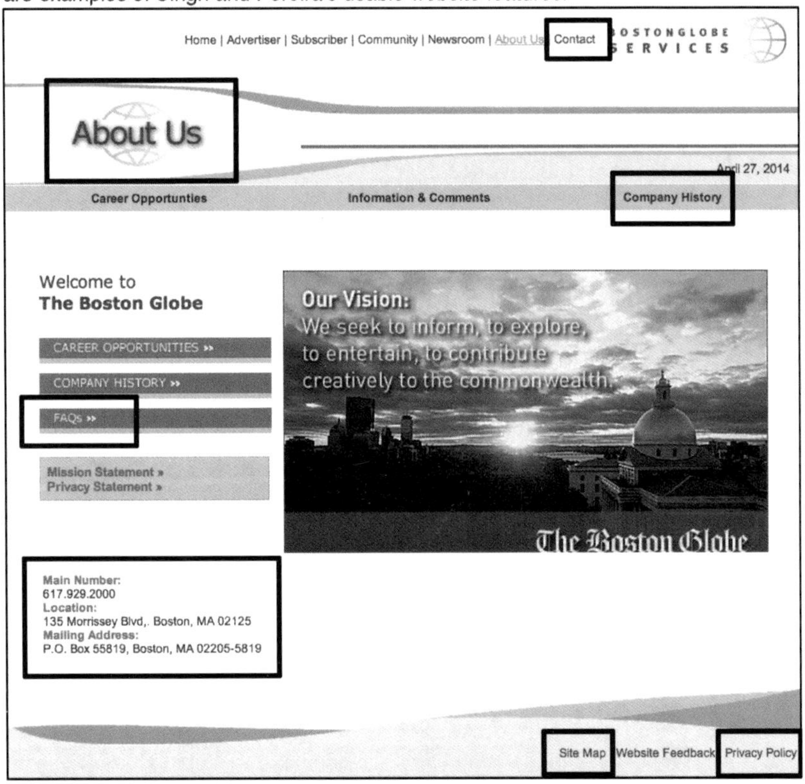

A basic knowledge of web design best practices is the logical starting point. The usability chapter in *Visualizing the Web: Evaluating Online Design from a Visual Communication Perspective* (2010) outlines several methods for both design and evaluation of websites. The basic starting point is the same—three questions that establish the mission, context and audience for the website under design or evaluation:

1) Who are the users? (there may be more than one user profile)
2) What is it they want to do or know?
3) What can/does the site provide so the user can accomplish their goals? (O'Connell, 2010, p. 125)

One should be able to determine what the primary function or strategic intent of a site is from the information provided in its content areas and navigation. Is it commerce, information, or entertainment? How do you know? Is it through the home page content or the navigation? Is there an "About Us" page that explicitly states what the strategic intent of the website is? Does that page give you clues as to who the site owners believe is the target audience? Is the target user someone from that culture or an outsider to the culture—a tourist? Having a description of the target user on the website jump starts the creation of a 'persona' or user profile, a brief dossier of characteristics that aggregate the user's needs and desires in reference to the website and beyond (see Creating a Persona in Chapter 3). Creating a persona for each user type—a technique first postulated by Microsoft designer Alan Cooper in his book *The Inmates Are Running the Asylum: Why High Tech Products Drive Us Crazy and How to Restore the Sanity* (2004), and widely adopted by design teams—helps keep designers from being self-referencing. Instead of thinking, "I'd like that!" designers ask, "What would Joe (or Helen or Kris) want?" Researchers can benefit from creating personas as well, collecting characteristics they find through the patterns and comparing these characteristics to the cultural profiles created by Hall, Hofstede, Rokeach, and others.

To thoroughly evaluate a website for its usability in reference to uncertainty avoidance, one must role play the target user and explore the site's navigation and interactive processes, identifying the information threads and work flows for clarity of purpose and design. It helps to have a few task scenarios based on both the profile of the target user and the strategic intent of the website. If, for instance, the website is for a local childcare facility targeted at local parents, what would be the primary task your target user would like to complete at that site? What, based on your assessment of the target user's culture, would be the concerns, expectations, values, and attitudes, they would bring to accomplishing the task? Armed with this knowledge (which should be captured in the user's persona) you could begin to explore the website with the user's needs in mind. Did you click on a link and find something unexpected happened? Would that bother your target user? Can they recover and try a different link? Would they? Or would they go to another website instead? Can they find the information they need to complete the task?

As I have already mentioned, what you see on a webpage falls into three categories, navigation, information and decoration. Decoration has its purpose, particularly in branding. However, unless it is masquerading as either navigation or process execution—for instance, having the 3-D drop shadow of a button—visual elements that are decorative generally don't pose a usability problem. You will have identified this situation easily in an Affordance test. Within navigation we have links that bring us somewhere and buttons that do something (interaction). Within information we have relatively constant or static information, such as contact information or historical or legal information, and dynamic information, such as news or computations based on information provided by the user. Advertising falls somewhere in between.

Navigation

If you did the Affordance exercise at the end of Chapter 7, you will have a grasp on the navigation conventions used by websites targeted at the culture you are researching. Navigation usually takes one of two forms: persistent and variable. Persistent or common navigation refers to navigational menus that show up on every page in the website. They also occupy the same space on every page, usually across the top or down the right or left side. Variable or contextual navigation is tied to the content on the page and usually leads to additional or related information. Occasionally, navigational links embedded in the content lead the user, not to additional information, but rather advertising sponsors. These are called interstitial links as they masquerade as variable content links but often lead to marketing or purchasing opportunities. Reputable websites label these navigational links as advertising so that the user can choose to follow or disregard them. Someone from a culture with high uncertainty avoidance would be leery of any site that didn't make clear what is advertising and what is editorial content.

An important point regarding navigation that is specifically cultural is the orientation and language of the navigation areas. Users expect that the alphabet and writing direction of a website targeted to them will be culturally appropriate. For instance, Arabic cultures would expect to have a site's navigation, as well as the site's content, read from right to left, with the most important or frequently used links also ordered from right to left (Barber and Badre, 1998). Cultures that use a specific alphabet expect the navigation and content to also use that alphabet.

Additionally, users expect the voice and idioms used in the language of the site to also be that of their culture. For instance, in English language sites, the home page is referred to as "Home." However, in French language sites, the home page is referred to as "Accueil"—

variously translated as 'welcome' or 'reception'—and not "maison." Danish sites uses "Forsiden" which is translated as 'home' but websites in Spain use "Portada" or 'front.' Oddly, German language sites use the term "Home" for their home pages. One can speculate as to the reason cultures use these metaphors, however there is a practical concern here. With so much machine translation happening on the web, much of it idiomatically incorrect, a multinational company seeking to localize their web presence will not be able to simply use a translation application if it wishes to accurately label the most important navigational link on their website.

Certain navigation conventions were set in place when the World Wide Web was first created in the early 1990s and these are still relevant today:

- All linked text should be colored and underlined.
- Links that have been visited before should appear in a color different from unvisited links.
- Links and buttons should be used appropriately. Links bring you to another place—either another page or another spot on the current page. Buttons execute an action—clicking on one logs you in or out of a secured area, adds something to shipping cart, makes a purchase, or submits data from a form. (O'Connell, 2010, p. 133)

However, there are exceptions to the above. When menu items are represented by icons, small simplified graphics, they might look like buttons even though they are actually links. To use a simple analogy, a door handle and a door lock do different things. The door handle, like a web link, helps you get to another place—pass through from room to room. The door lock, like the web button, executes an action. It secures the door and prevents it from moving. In many cases, the door handle and the door lock are separate mechanisms. But sometimes, they are combined. The same fixture does one thing or the other, as I learned to my chagrin the other day when I used the door handle to open and close the door, only to have locked myself out of my house because I didn't notice that the lock in the handle had been set. One might call this a usability failure. The same issue can happen in web navigation when the button-like image doesn't adequately convey its purpose as a link—or vice versa.

To further complicate the guidelines above, the same menu with the navigational button might include actual process icon buttons as well, e.g., print, post to Facebook or Twitter, email, or search. Using an icon is very powerful in that it metaphorically uses visual sensory information that transcends language and is easily understood in a global context. But the metaphor has to be unambiguous and the image cross-culturally understood. The *navigational* icon you click on in a menu should clearly indicate that its function is not to execute a *process* but rather to bring you to another place. The most typical icon of this type to be found in a menu is the "Home" icon, sometimes represented as a little building or house.

Interactive Processes

Getting lost in the navigation of a website might be frustrating but nothing generates quite as much anxiety than getting into the funnel of an interactive process with no idea as to the outcome of clicking on that button. Ask anyone who has inadvertently clicked on a button that was pre-filled with their email address in any one of the social media sites we, sometimes of necessity, frequent: LinkedIn, Facebook, Twitter. All try to get you to contact everyone in your email application's address book to build their list of users. With one little click, everyone who has ever emailed you gets an invitation to friend you or link to you or follow your Twitter feed. A moment's hesitation with that button could save you weeks of embarrassment. However, those in cultures with low uncertainty avoidance are less likely to hesitate and question, "What will happen when I click on this button?" A more *usable* website would provide you with an "Are you sure you want to do this?" pop-up.

In examining the interactive components of a website, we should put ourselves in a high uncertainty avoidance frame of mind and ask: Do the forms and processes allow users to back up or must they abandon the process to make changes? Are labels clear and unambiguous? Can the user find definitions for unfamiliar terms? For instance the U.S. Internal Revenue Service website let's you click on blue underlined links to see the definitions of important terms. Does the site use widely accepted conventions and follow established usability guidelines? How severe is the consequence of clicking on the button? When you do click on a button and the resulting action is irreversible, is the user given a chance to quit the process or go back to correct information?

An important aspect to websites seeking to allay the concerns of high uncertainty avoidance users is the ability to provide important information in such a way that the progress through the process funnel remains intact. For example, if the desired goal of the website is for the user to make a purchase, but the user may want additional information about the specifications or the product well into the purchase funnel, the web designer will have to make it possible for the user to access information without abandoning the process. Links that open up a new window with the desired information would be one way to handle this situation. Another, but perhaps less desirable, method would be to allow the user to save the process and return to it. Government sites with long, complex application forms for things like visas or student financial aid use both methods. At the end of every page in the form process there is a save button. Every part of the form that uses terms or calculations that might be unfamiliar to the user has links that open pop-up windows with definitions and explanations. The more uncomfortable the user is with ambiguity or uncertainty, the more you will see these accommodations embedded in the interactive components of the website.

Information Patterns

The information areas of a website fall into at least three different categories: relatively constant information; dynamically changing information; and monetizing information, either direct or indirect marketing and advertising intended to support the financing of the website. A usable website makes it clear and unambiguous which category the information falls in by the typography, location and labeling it uses for the content. Additionally, it is the combinations of these information categories that create patterns you can identify based on a website's strategic purpose.

Many of the visual cues that help us understand what is dynamic or constant content, editorial or advertising have to do with typography, the visual characteristic like color, shape, size, orientation, spacing, even texture. Letters that are heavy and larger than the rest of the content are interpreted as entry points: headings or headlines that indicate the nature of the content that follows. Large blocks of text that are left or right justified, depending on the language of the culture, and evenly spaced are taken to be body content. These are conventions borrowed from print media and were established online at the very birth of the web. The HTML or XML code that underlies a webpage basically controls the typographical and image elements we see on the page: the fonts, spacing and layout that we see when we look at a webpage.

HTML, or Hypertext Markup Language, does more than just determine what the words and images will look like when rendered in a web browser. The code also determines how that material will behave and what one can do with it. For instance the code <H1> is a first level heading. Words between those tags will be larger and heavier in appearance than the body copy between the paragraph tags, <P></P>. When search engines index the material that falls between the <H1> and</H1> tags, those words may score higher than other words on the same page if they match the searched for keywords. Why? Because the terms within these heading tags presumably are more relevant to the content on the page than random words that would be between the paragraph tags. Of course, it is not at all that simple and search engines like Google and Bing! have complex algorithms that predict relevancy, frequency, and reliability of terms found so that you stand a good chance of finding what you are looking for when you use their search engines.

In addition to the three different kinds of information, constant, dynamic and monetizing, these information types combine to form patterns: dynamic information driven sites and constant information driven sites. Monetizing sites fall into one or the other pattern depending on the amount of automation underlying what you see on the page.

First, in order to better understand the categories of information commonly seen on the web, let us look specifically at a simple webpage, the search results page of a Google search looking for air conditioners

Figure 9.2.1 Google search for small air conditioners

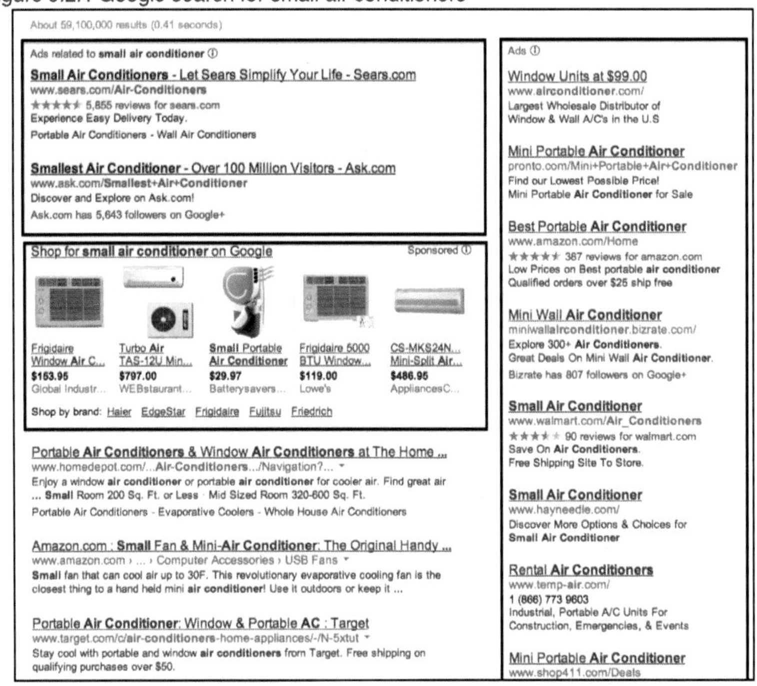

Figure 9.2.2 Google page static information

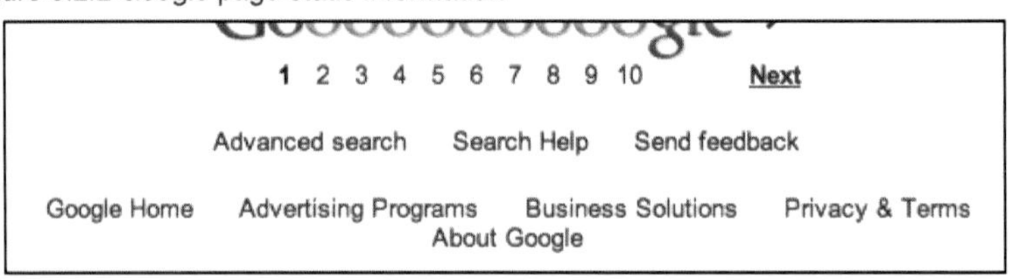

(Figure 9.1). Everything on this webpage is dynamically generated. There is no constant information at all. The boxed areas are three different areas of monetizing information, and each area is clearly marked as advertising. The three search results visible in this image under the sponsored results are organic search results that best match the Google's algorithm for the search terms "small air conditioners" and are, not surprisingly, from three very popular merchandizing websites for home goods: HomeDepot.com, Amazon.com and Target.com. Google's reputation as a reliable search engine depends on the user being able to know when the results of the search are organic or paid for advertising.

There is no constant information on this page. At the bottom of the search page are links to that kind of information: privacy and terms of use information, a page about Google, information on their business products and advertising programs (Figure 9.2). The information on these pages doesn't change much but it is important to Google's users when they need or want it. The excellent usability of the Google site makes it possible for even the highest uncertainty avoiding culture to have confidence in using Google and possibly contributes to its almost universal popularity. It is ranked by Alexa.com, an Amazon.com website, as #1 in the United States. It is ranked globally as #2, outranked only by Facebook.

News sites are also driven by dynamic information with many monetizing information areas, as we can see by looking at the *Wall Street Journal* (Figure 9.3). Again, I have boxed the monetizing information areas, but unlike Google, monetization on this page is split between advertising and subscription based access—the little key icon indicates content that you cannot access without subscribing and logging in. All the content on this page is dynamic in that it changes daily, if not hourly. Some of the content, like the stock exchange information in the upper right corner, might even change every few minutes. There is ample evidence of headlining, each headline a live link to the full story, and teaser copy—enough copy from the story to let you decide whether you want to drill down to the full story or not. In the Opinion block, there is only the name of the columnist and the title of the article.

Just as with Google.com, there is no constant information content on this page. The bottom of the page provides links to the constant content: policies and terms of use, information on advertising and customer service information, although, if you click through to the Customer Center, the footer displays a slightly different set of links that include a link to "About Dow Jones", the parent company, and another link to a Site Map (Figure 9.3).

Tool sites like search engines and news sites like the Wall Street Journal, the BBC and others that provide dynamic consumable information, attract a great deal of daily, even hourly traffic. In contrast

Figure 9.3 Dynamic content the WSJ uses to monetize the site and the constant footer information displayed at the bottom of each page.

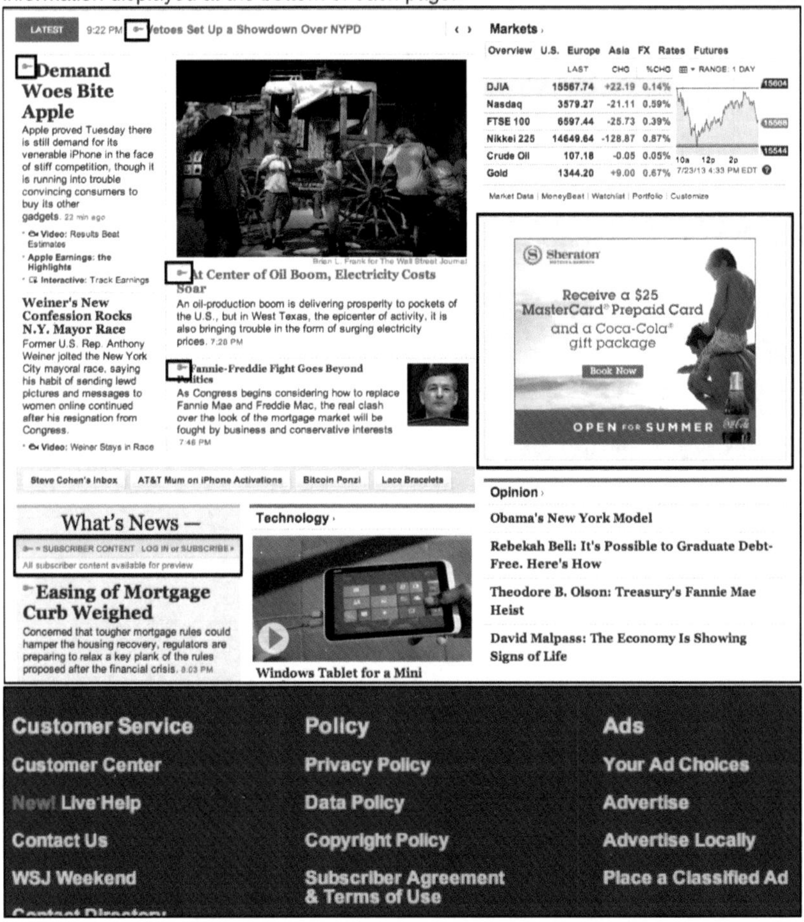

the vast majority of general information webpages like government sites and product marketing sites rely on relatively constant information, although some of that information may be animated. Because the algorithms used by Google and others include a popularity quotient and popularity is a result of visits and links, these sites have found it a challenge to provide the kind of dynamic consumable information that generates the levels of traffic required to stay at the top of search engine results. In recent years, social media and blogging have solved the problem for these sites, establishing an environment where customers and citizens can 'converse' online, helping to create the very content that keeps parts of a company or service website fresh with new dynamic content. However, because the strategic intent of these sites is to provide relatively stable, constant information, the dynamic information areas are a small percentage of the overall information landscape. Take the New Alexandria Library as an example (Figure 9.4):

Figure 9.4 New Alexandria Library website

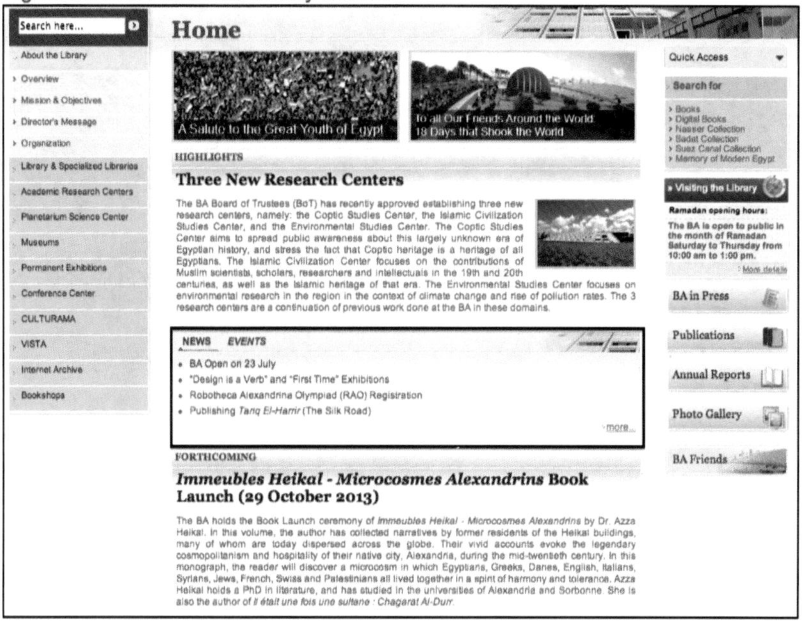

Less than a quarter of the library's home page has dynamic content and even that is only a list of headlines. The rest of the information content consists of stories with significant 'shelf life' and links to information that likely does not change at all, such as "About the Library," "Mission and Objectives" and other information about the institution. Unlike the Google and Wall Street Journal sites, the links to this constant information is in the upper quadrants of the home page or 'above the fold,' a metaphor borrowed from print newspapers that indicates the most important part of the page. You will also note that there is no advertising or monetizing on this site. That is not part of its strategic intent.

Let us look at a product merchandizing site's home page. This is the home page for the Toyota Prius (Figure 9.5). Not only is there no dynamic information on this page, nearly all the information that is there is visual. There is some user-driven animation in that one can click on one of the color circles under the car and see it in a different color, but that is not dynamic 'information'. Several days from now, perhaps even months from now, the information concerning this car will be the same—until the

Figure 9.5 Toyota Prius website home page

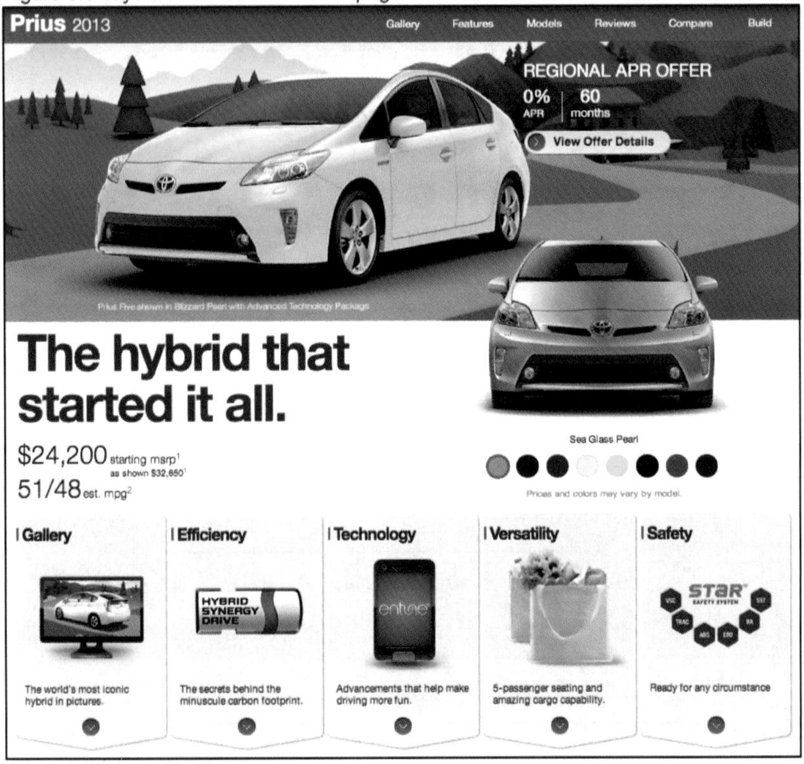

2014 Prius comes out. There is also, arguably, no monetizing information as the strategic purpose of the site is to get the user to a dealer. There are no offers for other products to compete with this one goal—and the user can easily drill down to see even more constant information.

There are many information patterns to be found on the web. Example 4 in Chapter 10 is a discussion of news websites whose bricks and mortar corollary is the broadsheet newspaper. Newspapers have long monetized their information product by providing advertising space at both the corporate ("classifieds") and the personal level. Indeed, for over a century, the personal ads were a key driver of circulation. In moving to the Web, newspapers failed to migrate this critical revenue generator as websites that served the same functions—online dating, lost and found, and sites like eBay and Craig's List—had already established successful Web presences. However, one can still see vestiges of the traditional newspaper in the look and feel of a newspaper website: the layout, the news desks and, yes, the classified ads.

Summary

The information content of any website is, hopefully, where the user spends most of his or her time. A usable website allows a user to come to the site, find what they need and then move on. For users from cultures with low tolerances for ambiguity, confusion or uncertainty, a usable website is where:

- the navigation is unambiguous,
- the consequences of the interactive processes are predictable and well annotated,
- the information content is easy to find, and where
- the ability to drill down to more specific, richer information is logical and easy to navigate.

For these cultures, usability generates the highest levels of confidence. For the rest of us, usability just makes the whole web experience more pleasant and less frustrating.

Basic Funnel Flowchart

In the kitchen, a funnel is wide-mouth tool that lets you pour things into containers with much smaller openings. In web-parlance, a funnel is the process you put a user through when you want to narrow their options and reach a desired goal—like a purchase or sign-up. There are several funnel tests a designer/marketer can conduct. Most involve increasing the success of the 'funnel' in keeping users from abandoning the process. However, we are more concerned with observing the ambiguity and uncertainty some funnel processes may generate and in determining how the uncertainty avoidance tolerances of our target culture may respond to them. The best way to find possible trouble spots in a process is to map it out using a flowchart. Flowcharts were developed by Frank Gilbreth, the 'motion study' pioneer who introduced a scientific approach to manufacturing. Its power is founded in the standard use of a set of shapes to indicate function and the ability to use arrows to indicate sequence of actions and re-use of steps in the process. These are the shapes and connectors (Figure 9.6):

The steps for creating any flowchart are as follows:

Figure 9.6 Flowchart symbols and connectors

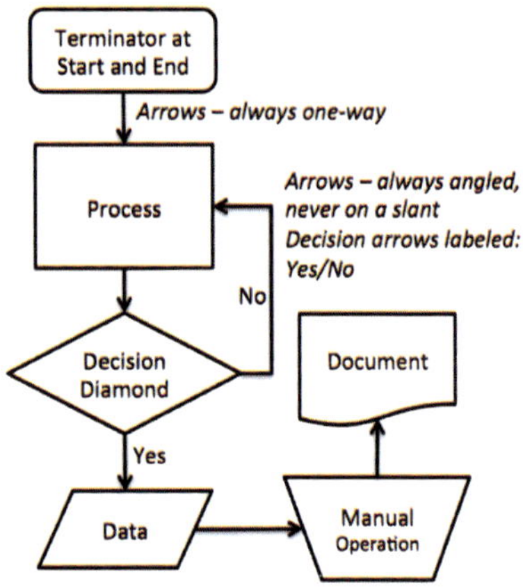

1. **Gather** information on how the process you are mapping flows.
2. **Define** the boundaries of the process clearly so that you don't get 'scope creep'.
3. **Map out** the flow using the appropriate symbols —use the simplest symbols possible for the given step in the process.
4. Make sure there is only one output arrow out of a process flow. If there is more than one, then you have a decision in the process.
5. Frame the decision so that it only has a "Yes" or "No" answer. More than one of either indicates an additional step in the decision.
6. Make sure every feedback loop (re-use of a step) has an escape. This usually requires a decision diamond somewhere in the process.

When you have completed your flowchart, run through the flowchart with someone else to see if you have captured all the steps and have mapped the logic of the process accurately.

Purpose: To evaluate the usability of online processes and forms.

Method: Pretend that you are on the design team for Amazon.com (or some other web book or product merchant) working to improve the order process at the site. First you must document the current flow or 'order funnel.'

Create a flowchart for a NEW USER (no account) that outlines the process of buying one book on Amazon. Please capture the decision points with the appropriate decision diamonds. REMEMBER to use the appropriate symbols and to label the Yes/No lines from each decision point consistently.

Remember to simplify the representation of the process by 're-using' steps.

Capture all the possible endpoints.

Highlight in yellow any possible points where a new user can become confused or stuck. Indicate if there are 'help' links or prompts that help the user move on.

Remember, the object of the funnel is to get you quickly and completely to the desired endpoint. Any place in the funnel where you might get frustrated and abandon the process is a trouble spot. Any place in the funnel that gives you an option that drags you away from the desired endpoint is also a trouble spot.

Determine if the trouble spots are minor, moderate or serious. Even one moderate or serious trouble spot will be a problem for a culture that has low tolerance for uncertainty or ambiguity, or high uncertainty avoidance.

Taken together, the usability of each process on a particular website and the attendant help features like pop-up definitions, links to customer service, FAQs or Frequently Asked Question pages, all create a pattern of that site's overall usability and predicts how a user will perceive the 'match' of the site to the user's tolerance for uncertainty and ambiguity.

PART 3: PRACTICAL APPLICATIONS

As the approach in this book is largely 'how to' use a pattern language to conduct examinations of websites to discover their embedded cultural aesthetic, some illustrations of how the data is collected and documented in the cultural grid is helpful. This section provides examples or 'case studies' of cultural aesthetic research so the reader may have some models for pulling the information together for his or her own project. Both geo-ethnic and functional patterns are illustrated as well as templating phenomena observed in websites created by multinational corporations. My students, upon completing their research, were asked to give a 'tag line' name for what they were seeing. The names for the patterns used here are not indicative of anything more structural than what they reveal about the researcher's overall impressions. The following examples are included:

Regional culture:
Denmark – "Scandinavian Design"

> This pattern has been a component in the *Visual Media in a Cultural Context* course since its inception. It is a fairly stable pattern revealing a great deal of cultural longevity as, year after year, the same pattern elements show up. This example illustrates how the cultural grid, examples and notes work together to describe a cultural aesthetic pattern.

Diasporal culture:
Argentina – "Pictures and People"

> This culture's pattern has also been a component in the *Visual Media in a Cultural Context* course since its inception. One year, I had a student who had gone to Argentina for a semester study abroad. She was able to validate her fellow students' perceptions based on her experiences living with a family in a small town outside Buenos Aires. Students were encouraged to look at sites from Spain, Italy and Greece as well—to compare and contrast old world and new world cultural patterns.

Multinational templates:
Starbucks & Coca-Cola - Localizing "Web Flavors"

> At one point, large multinational corporations—at least those that cared to—spent a great deal of money localizing websites for their customers in other parts of the world, perhaps to gain an edge over competitors who didn't do much more than some automated translation into a few official languages like French, Spanish and German. But this became an expensive way to do business. Enter the culturally modified "template set." This example shows the different templates designers in two large multinational (some might say ubiquitous) brands used to 'localize' websites. There is certainly more research that can be done in this area. Thanks to the Web Archive's "Wayback Machine" we can go back in time and look at websites that have been archived since 1996 and compare them to the current designs. Not all the graphics show up and, just like any research project, scanning through the web crawl results is time consuming, and sometimes frustrating and tedious, but it does mean that the web isn't quite as ephemeral as we sometimes think.

Functional templates:
News websites - "Print goes online"

> Most media enterprise sites appear to have the same visual structure as their physical-world counterparts. Television network websites (BBC, HBO, CBS), have moved to adopting a programme guide structure. Visits to the Wayback Machine indicate that in the early 2000s, this was not commonly so. TV sites, at that time, tended to look just like most of the other corporate websites of that era. Interestingly, America's premiere programming guide, *TV Guide*'s website looks like a traditional news website. In the case of newspaper sites, the *New York Times* has sported the same overall organizational structure since the late 1990s, one based on its print newspaper. Its British cousin, *The Times*, looks very much the same. Other news sites seem to base their look and feel on the paper versions of their own publications. For instance, *The Guardian*'s website has a tabloid feel to it, as does thesun.co.uk. An exploration as to why this is the case would make for a great research project.

Possible applications of this pattern language and the research methods that support it for different users are also reiterated here.

In the *Afterword*, these examples and their attendant observations conclude with some personal thoughts on the future of this pattern language, its ability to shift and change over time just as other languages do. It is also a philosophical rumination on the possible effects of globalization on culture and cultural expression. This is an area of cultural studies that is deeply impacted by modern digital technologies as digital applications such as the web and social media make national boundaries permeable and reinforce the migration of ideas while, perhaps, lessening the demand for physical migration in terms of human capital. The final chapter is, however, not the last word. It is an invitation to further exploration and expansion of the pattern language itself. As Web 2.0 technologies proliferate, will cultural aesthetics disappear as templates homogenize the visual landscape of the Web? As new technologies and platforms emerge that are more visual, how will cultures that are verbally oriented respond? Will we see a shift in the privileging of words or images? Many more questions and suggestions for further exploration will likely emerge and I hope we can continue the conversation about cultural aesthetics we have begun here.

Chapter 10

Using the Pattern Language

At this point it may be useful to suggest how our pattern language can most profitably be used to map the aesthetics of a culture. In this chapter, I provide a recap of the pattern language, discuss various approaches to its use and follow up with several examples, or case studies, that demonstrate how the pattern language can be used to collect and investigate a culture's visual artifacts and determine what they tell us about how a culture is represented in various visual media.

Pattern Language and Tools

To briefly recap, our pattern language consists of dominant pattern elements such as color palette, content, information flow, time mode and tempo, and cultural beliefs, attitudes and orientations. Each of these has been explained in terms of both its provenance in the fields of cultural anthropology, technology or design, and each is listed in the Cultural Aesthetic Grid wherein we can index our findings for each website and visual artifact we study.

A specific research project can start by either identifying what we believe is the target user before looking for websites that meet their needs/task criteria—this would be a usability design approach. Or, we can begin by looking for websites that appear to be created for a particular culture and deduce, from what we see, the 'persona' of the target culture's typical user. In either case, we should end up with 40-50 websites that have been created by and for the target culture. At this point some dominant themes should begin to emerge. These are that specific culture's aesthetic patterns and an understanding of these should help us better understand what visually resonates with that community.

Applications of this Knowledge

There are perhaps as many applications of this understanding of a culture's aesthetic patterns as there are cultures to research. However, at the outset of writing this book, I identified specific audiences who might find this methodology useful.

Educators and students of intercultural and visual communication may use these methods in a course similar to the one that inspired it. When I teach this methods course, I frame it as a 'journey without,' an exploration of cultures that are often far removed from our personal experience. However, I also know that to fully engage in a 'journey without,' one needs to experience a 'journey within.' As such, it is as important to explore one's own cultural aesthetic milieu and examine its underlying beliefs, values and behaviors, as it is to try to identify and understand what we are seeing in the artifacts of other cultures. The tools used in this pattern language work just as well when examining one's own cultural aesthetics.

Scholars and researchers in visual communication and design can benefit from using a consistent pattern language as a basis for cataloging not only cultural aesthetics but also shifts in technology/design cycles. For example, I started to notice various technological shifts as, over the years, each new crop of student researchers began to plow the same Web territory. In 2004, many of the sites we studied were relatively low-tech in their designs. The communication network resources required to support bandwidth-greedy Flash™ animation existed only in advanced technology first world countries with the infrastructure needed to support high-speed downloads. As more and more countries invested in that infrastructure, sometimes leapfrogging telecommunication technologies, we began to see a homogenization of aesthetics. Website developers in these countries were using templates created by the big web software companies (Microsoft and Adobe) to rapidly get these high-tech looking sites up and running. Eventually, the fad aspect of technology gave way to the dominant cultural aesthetic and we began to see cultural customization again—until the next technology wave came in. This has evolved into what I call the technology/design cycle wherein there is a development learning curve embedded in each new technology adoption with a subsequent reassertion of the cultural aesthetic when sufficient mastery is achieved. When we see imagery or layout designs that seem inconsistent with what we know of a culture's aesthetic preferences, we might speculate that perhaps a new technological/design cycle is at play.

Communicators who rely on the Web to reach their target audiences must pay close attention to the cultural aesthetics of those audiences. The essence of effective communication is the ability to "speak" to the target audience's "listening." The better one can profile the modes of communication that are privileged in a given culture, the better one can shape the message to be heard or seen, read or viewed. Using this pattern language, or perhaps adding to it, can help communicators create more culturally responsive messages regardless of communication channel.

Creators of websites and other media that must convey cross-cultural or intercultural messages can benefit from a dictionary of these patterns. Following cultural aesthetic patterns helps improve communication—it helps the targeted communities 'feel at home' with the visual landscape. However, it doesn't work by itself. Familiarity with cultural values and taboos is essential. Cultural aesthetic patterns misapplied will generate cognitive dissonance and a sense of unease. Nevertheless, a cultural pattern language is a valuable tool for designers of visual media.

Using the Patterns and the Grid

This chapter gives several examples for using a pattern language method to identify basic cultural patterns and extrapolate a profile for producing more focused visual communication or for identifying emerging cultural trends. Much of the material used in these examples was culled from research spanning the last ten years. However, it is not the intention of this book, or this chapter, to codify or suggest that conclusions regarding certain cultural patterns are fixed. Rather, it is the work of this chapter to show the reader *how* to use this pattern-based method to do one's own investigating and exploring. The examples listed here are for demonstration purposes only and should not be taken as the determination of anyone's research efforts. Any visual analysis and heuristic evaluations outlined here were done in the context of a classroom. Coincidentally, these explorations into a culture's visual aesthetic also served to test and refine our pattern language.

In the examples that follow, we can identify familiar cultural aesthetics like "Scandinavian design." Creative director Rebekka Bay, in an interview for Fashionising.com says that this design aesthetic is about:

> … functionalism and minimalism, and taking away what isn't needed. "When you go to design school in Denmark they teach you to take things away, rather than adding embellishments and decorations, they try to teach you how to do with very little." The point, she adds, "is not to be borderline boring, but to focus on what's really, really important." (n.p.)

A design movement that has been defined by simple, uncomplicated designs, and low cost mass production, "Scandinavian design" appeared also to be an appropriate appellation for the patterns we observed as we analyzed over 100 websites from Norway, Sweden and Denmark.

Among these case study examples, we also look at a diasporal culture representative of high-context, polychronic preferences with a warm color palette—Argentina. Communities that are diasporal in nature carry the cultural values, attitudes and beliefs of their homeland to their new homes along with their material goods. Ethnically identified mediascapes, as Arjun Appadurai explains, "tend to be...narrative-based accounts of strips of reality, and what they offer to those who experience and transform them is a series of elements...out of which scripts can be formed of imagined lives, their own, as well as those of others living in other places" (1996, p. 35). Scripts are the story telling equivalent of visual patterns. In looking at the websites of colonized communities, it is important to be aware of layered meanings and the evidence of, perhaps, several ethnic cultures in the one place, each informed by their own set of values, attitudes and beliefs.

Shifting the idea of culture away from geo-ethnic centered to corporate or functionality centered websites, we look at some technological trends in 'glocalization'—global or multi-national companies that have purportedly customized their Web designs for local audiences. This example includes an examination of the use of design templates—a phenomenon that would appear to create a level of aesthetic homogenization. While on the surface, templates appear to be economical, without a basic understanding of how the templated design elements will be perceived by the target audience, they potentially move the enterprise further away from culturally customized or localized design with, perhaps, unintended consequences.

Finally, we look at how the format for a particular trade or purpose still evidences culturally aesthetic differences. By analyzing newspaper sites, we can examine how online news organizations mimic their print counterparts and adapt the online environment to their readers' behaviors and preferences. Both broadsheet and tabloid sites are discussed, but the broadsheet model is indexed in the Cultural Aesthetic Grid accompanying this example. The tabloid model is indexed in the Argentina example using *Cronica.*

While the verbal analysis in each example is a synthesis of many observations, at the end of each example is a pattern grid for one of the sites used in the research. As a rule forty to fifty sites would have been reviewed and analyzed to find the common themes and dominant aesthetic patterns each time we revisited the subject cultures. Each would have had its own Cultural Aesthetic Grid. The analysis texts that are part of each case study are also examples of how someone might record their observations based on the patterns captured in the grids for different types of study. I hope you find these helpful.

Example 1: Denmark—Scandinavian Design

Figure 10.1 Comparison of Danish websites collected in 2007.

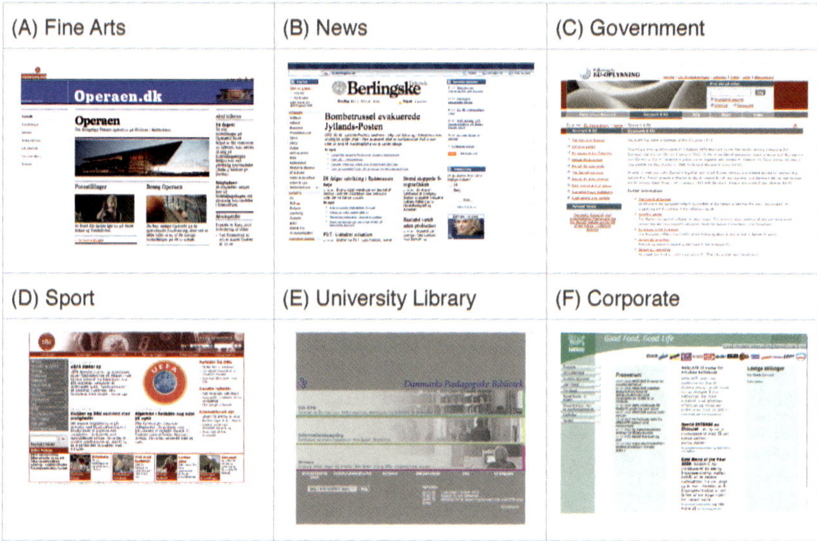

Analysis

Beginning with the color palette, with the exception of the national colors of red and white that we find in exhibits C and D (Figure 10.1), color schemes appear to be muted or neutral. The highly saturated or dark blues in sites A and B mastheads are cool, as are the reds in the mastheads of sites C and D, unlike the red in the UEFA logo in the information section of the page. While there appears to be a 50/50 split visual to verbal ratio on the Opera site (A), words appear to dominate. According to E. T. Hall, Hofstede and Wuertz, Scandinavian culture is low context and "communication occurs predominantly through explicit statements in text and speech" (Wuertz, 2005, Introduction). It appears that the content modality of this culture is verbal. This is illustrated in five of the examples above (A, B, C, D, F) as there is a good deal of verbal text on these home pages and even more text on the secondary pages. Example E is a 2007 screenshot for a university library home page. While this version of the site is light on words, the sub-pages of this site were quite text heavy. A 2013 screenshot of the same website's home page (Figure 10.2) shows that, while the same color palette is used, the visual/verbal ratio evidenced here is about 50/50 with a great deal of white space replacing the gray of the 2007 site.

Aesthetically, we notice in all the examples that, despite the amount of text, there is still ample white space so we would categorize 'content density' as low. The low content density is consistent with a desire for clear, uncluttered displays and distinct areas of categorization typical of Scandinavian design principles. Everything in its place and every place clean and functional.

Another indicator of this culture's verbal modality is evident in the navigation. Cultures with a preference for image modality will likely include images in a website's navigation. From the evidence on these example sites, it is fairly clear that all of the navigation in both the menus and contextual links is predominately, if not exclusively, verbal. This level of simplicity and clarity in the navigation supports a monochronic culture. The user is subliminally encouraged to pursue one task at a time. Each area provides a great deal of the information necessary to complete a task without going to other areas of the site. Because so much of the information is in verbal form, websites load very quickly, optimized for speed and efficiency. This, too, is consistent with observations made by Hall and Hall, Hofstede and Wuertz.

Audience interaction tends to be more static than dynamic in the majority of these sites, with the exception of sports sites like dbu.dk (Dansk Boldspil Union –Danish Football Association), where the 2013 site (Figure 10.3) features one video block and the ability to search Google Maps for the location of the next game. This video block and the Google Map application were not present in the site originally studied in 2007. The rest of the sites in this sample do not offer much interactivity and the information flow, while transparent and open, follows a logical branch structure. The user will not easily get lost on any of these sites. The information density is very high. There is plenty of information to drill down into.

According to Hofstede, Danes are individualists (p. 53), however, they are also team players, have strong social values and are committed to their families, friends and communities (Lewis). Unlike the American model of rugged individualism, or every man for himself, the Danish model is creative individualism supported by 'hygge,' a sense of community and the security that community relationships provide (Lewis, denmark.dk). They are willing to pay some of the highest taxes in the world in exchange for nationally funded healthcare and education.

The imagery of the sites that were collected in between 2004 and 2012 did not feature many people. What imagery there was consisted of inanimate objects or abstractions. The 2007 news site example, Berlingske (Figure 10.1 B), was 100% verbal in the information and navigation areas with only small images as teasers for the television division of this major

Figure 10.2 DPB Library 2007

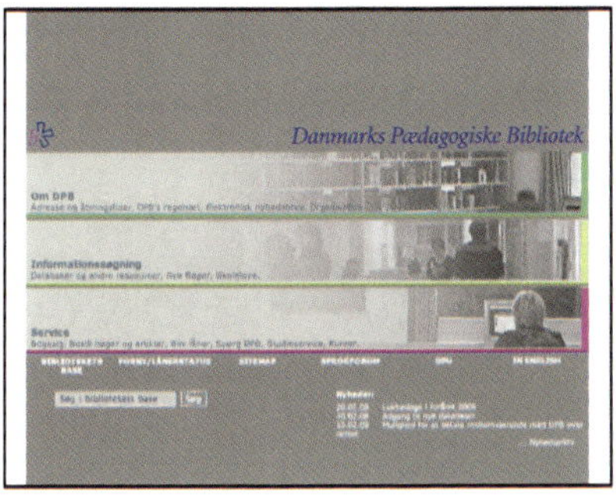

DPB Library 2013

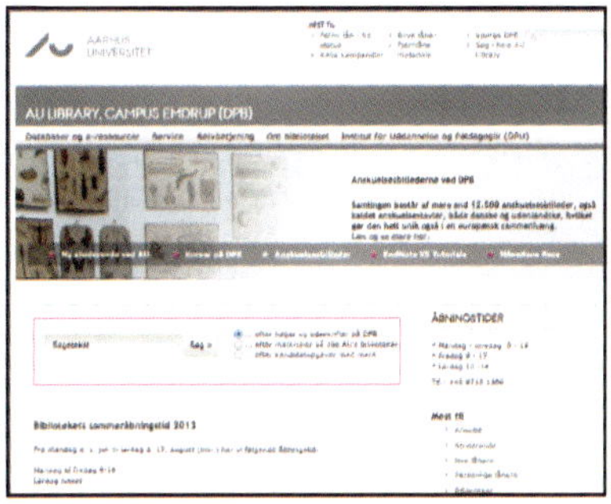

Figure 10.3 Berlingske News 2007

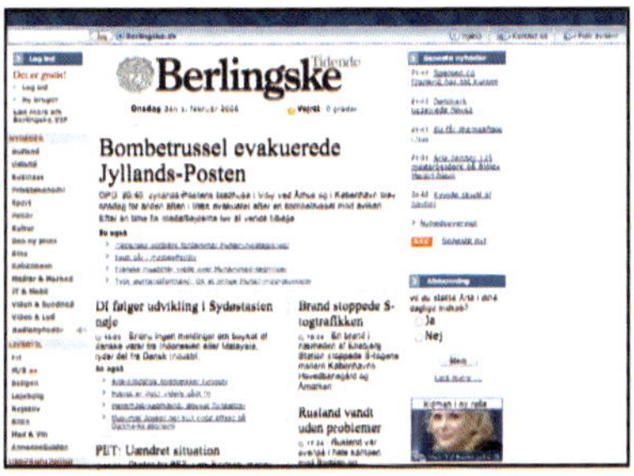

Berlingske News 2013

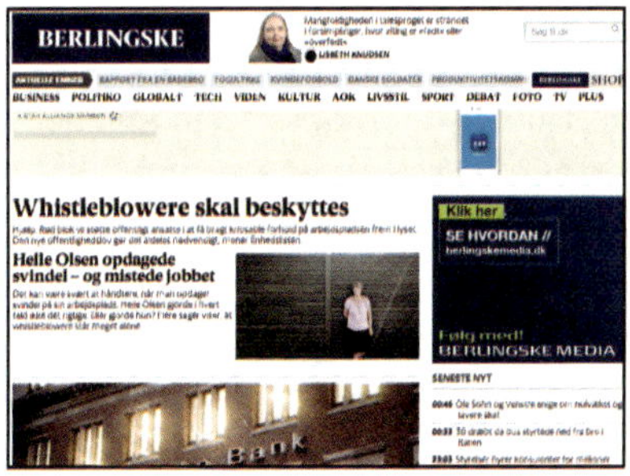

Figure 10.4 Danish Football Assoc. 2007

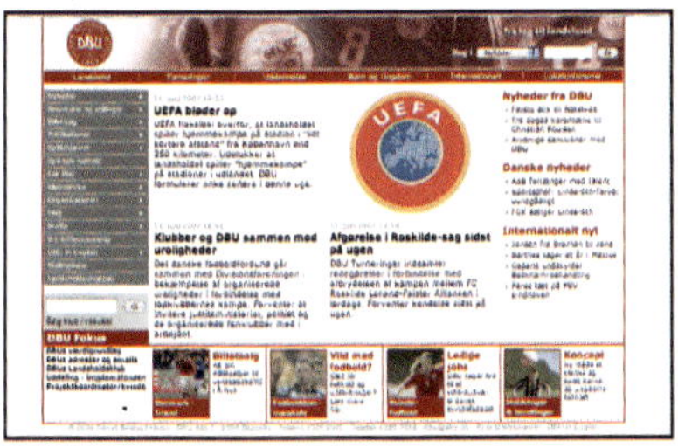

Danish Football Assoc. 2013

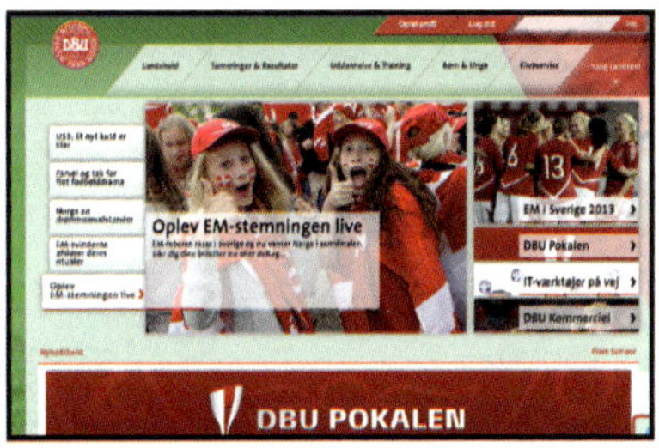

Figure 10.5 Erhvervsstyrelsen 2013 (Grid example)

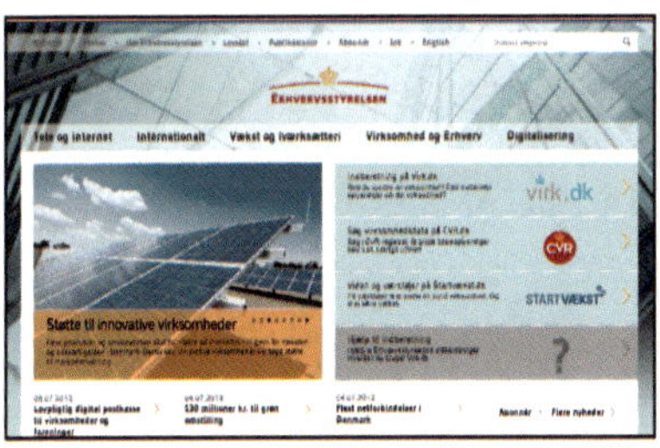

media enterprise. The 2013 version (Figure 10.3) features headshots of reporters and columnists in their masthead and more news images in the information sections. However the content is still dominated by the verbal mode. The imagery on the 2013 version of the Danish Football Association's site (Figure 10.4), in contrast with the 2007 site, is saturated with images of people. There is very little verbal content and most of what we see as words is either coded as headline or navigation. The site has also changed their color palette from red and white to bring in the green of the football pitch.

Using a 'thirds' grid to compare the 2007 version of the *Berlinske* website to comparable news sites in other parts of the world helps us more clearly see similarities in the information pattern while, at the same time, making explicit the visual verbal ratios and the modality of Danish website aesthetic patterns (Figure 10.6). Both India and Denmark show a preference for a verbal modality, whereas Iran and Argentina show modality preferences that are 50% or more visual with higher content density. Yet all these examples show the columnar layout typical of their respective newspaper layouts, be they broadsheet or tabloid.

Figure 10.6 Comparison of news websites using a 'thirds' grid.

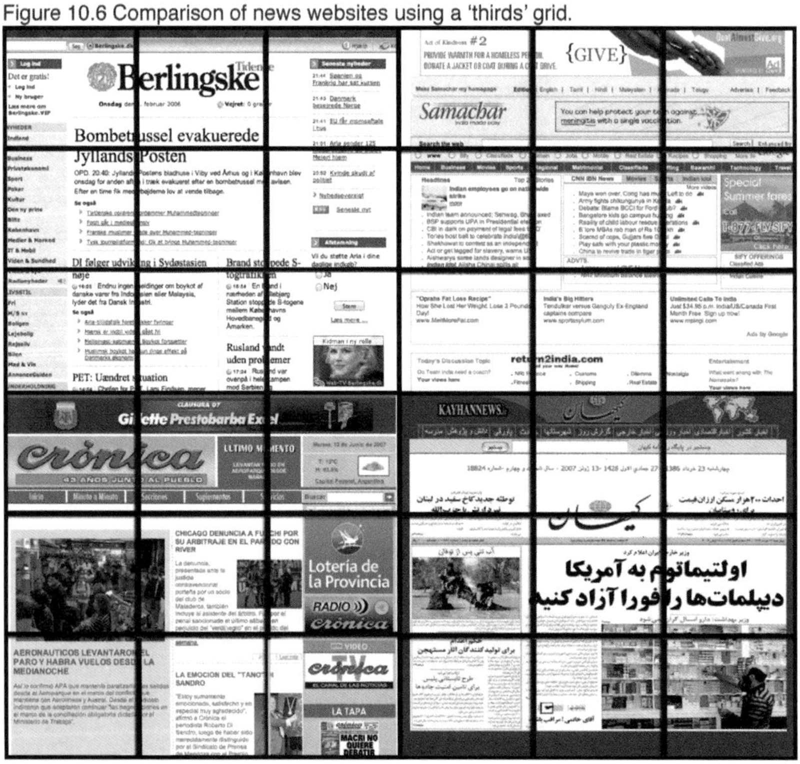

There are many more layout patterns that, because of their success in either the "dirt" world or the virtual world, have become archetypes for websites that seek to compete at some level in the same market area. In looking at Danish websites, those that did not conform to the observed cultural patterns were often imitating multinational competitors or were subsidiaries of multinational companies.

Research into other studies concerning the attitudes and value orientation of the Danish culture support the material observed on the sample websites. This is a culture that is rational, objective, but still focused on community and business relationships. Lewis writes, "Danish flexibility, tolerance and business acumen are three of their outstanding characteristics." They value "honesty, cleanliness, work ethic, egalitarianism, social justice, equality for women, and tidy public spaces" (p. 293). And they have a sense of irony and humor. Unlike other cultures in their geographic neighborhood, their general attitudes appear to indicate low uncertainty avoidance, and indeed, Denmark ranks third lowest in Hofstede's Uncertainty Avoidance index, just above Jamaica. However, Danish websites are so usable and well organized that we can only assume that their designers place a high value on usability because it is consistent with their aesthetic value of simplicity, clarity, and functionality.

In all the websites studied, lines are fairly straight and clean, the overall designs are uncluttered and unambiguous in their strategic intent. If we look at Scandinavian film, architecture and furniture design we will find this same simplicity and frugality—form follows function and what adornment is there is highly functional, as a quick trip to IKEA.com will attest.

Table 10.1 Cultural Aesthetic Grid - Denmark

Pattern Variable	http://www.erhvervsstyrelsen.dk/ (Figure 10.5) Government Business Development Bureau
Color Palette Warm/cool Saturated/Muted National/Natural Contextual/Brand	Cool or neutral colors Muted shades Natural palette
Content Modality (Image/Text) Ratio (% Visual, % Verbal) Density (High/Low)	Text appears to be more important 50/50 split except for government and bank sites Low density - lots of white space and clear organization
Information Flow Fast/Slow Poly-dimensional/Uni-dimensional	Fast on the home page. Short descriptors to get the user where they need to go. Appears to be polydimensional - transparent and open
Time – mode Polychronic/Monochronic	Monochronic
Time – tempo Fast/Slow	Very fast download speeds. Site appears to be optimized with much info in verbal form.
Audience Interaction Static/Dynamic Serial/Branched/Hypermedia Passive/Interactive	More static than dynamic Branched flow Interactive to a degree… especially in product sites (IKEA)
Information density High/Low	High… a lot of information, plenty to drill down into
Imagery Collectivist/Individualist (many people/one or two persons)	Often no people at all. Only in the one slideshow on the home page and few of those images have people. Mostly objects or abstractions.
Context High/Low	Low.
Attitude Rational/Emotional Objective/Relational	Rational, objective, but still focused on community and business relationships
Value Orientation Materialistic/Quality of Life	Quality of life is emphasized over materialism
Usability Ambiguity tolerant/intolerant	High usability, low ambiguity
Uncertainty Avoidance High/Low	Traditionally low, so this could be emphasis on usability and 'good' design

Example 2: Argentina—Pictures and People

Figure 10.7 Comparison of Argentine websites collected in 2007

(A) Bank	(B) Government	(C) News (2007)
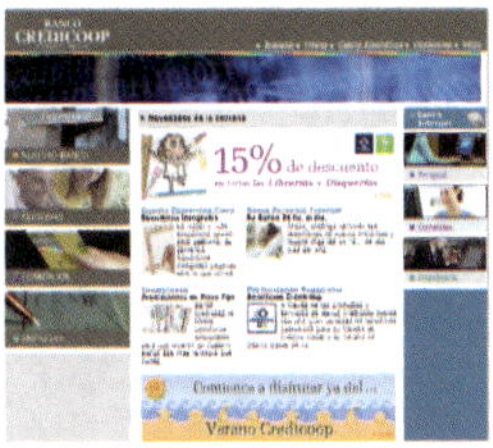	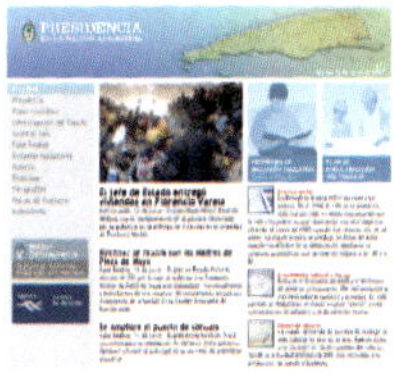	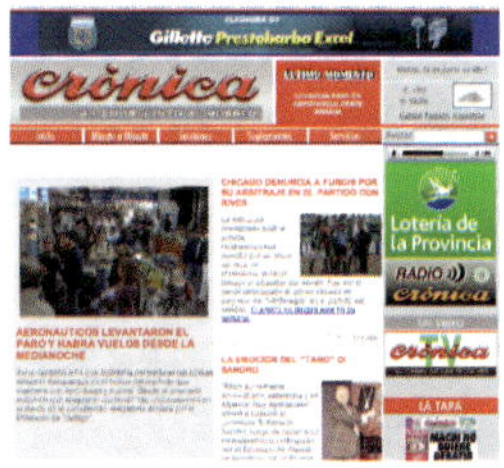
(D) Local Restaurant	**(E) Credit Card**	**(F) Dancing Lessons**
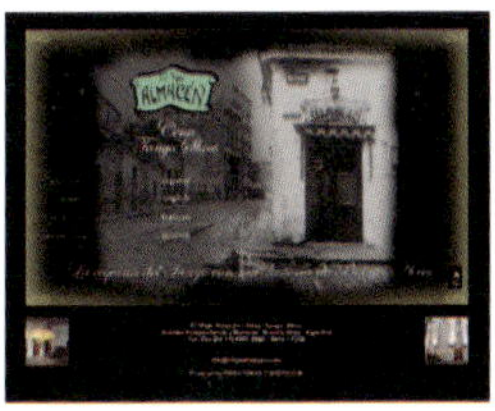	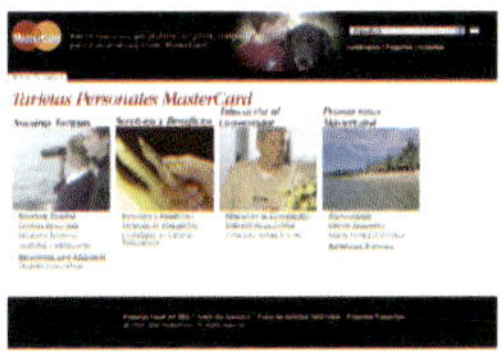	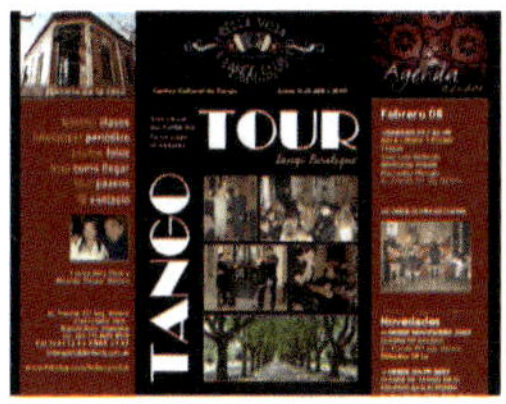

Analysis

Argentina, like many of the Latin cultures (Mexico, Italy, Brazil) is, socially, a high context culture with a strong emotional attitude that includes bold imagery and strong, warm colors. In contrast to northern European cultures like Denmark, the aesthetic is much closer to those found in the Mediterranean area and throughout Latin America, despite the fact that Argentina is almost the southernmost country in South America, less than 600 miles from Antarctica where winter reigns all year long. One may speculate that the preference for tropical "latin" color palettes (see Figure 10.8) stems from Argentina's colonization by Hispanic Western Europeans going back to the sixteenth century.

Figure 10.8 Caminito, Buenos Aires. Source: B. Krupski, 2007

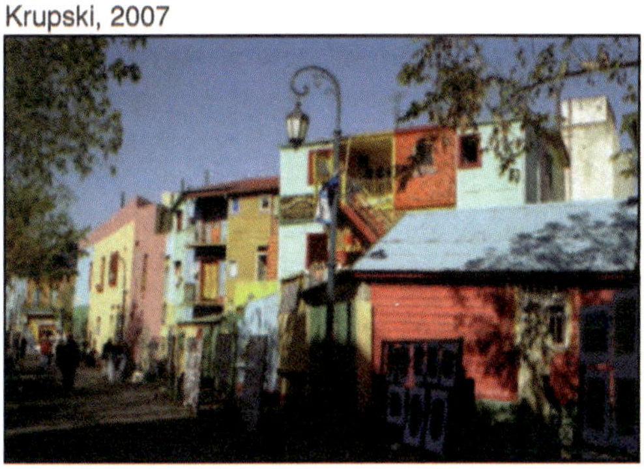

In terms of content density, we find that there is much less white space when compared to low density cultures like Scandinavia. The visual/verbal ratio on many sites, particularly media sites like Cronica.com.ar, leans toward images. *Cronica*'s more conservative news competitor, *La Nación,* exhibits a much more 50/50 split on their home page, a ratio that is more visible as one moves down below the fold and away from the big advertising blocks. Websites studied in 2005 (see LA NACION LINE Figure 10.9) and 2007 (sites in Figure 10.7 A, B, C, D, E and F) indicated a culture that looked for verbal information so this appears to be a drift toward more image modality, perhaps as a result of better technological infrastructure or, as can be seen in Japan's websphere, a split between the older verbally oriented culture and the younger media centered culture.

The information flow is fast as the information images are highly descriptive. Flow is also poly-dimensional with no obvious barriers to information access, and a great deal of 'drill down' in the finance, higher education and government sites. Most sites also accommodate comments and links to social media applications, something that would not have been present in earlier studies (2004 – 2010) of these sites. The site that is documented in the Cultural Aesthetic Grid below is for Cronica.com.ar—a tabloid news organization similar to *The Sun* in Great Britain. Hence, audience interaction observations for this site are skewed to a particular taste in journalism. In government, finance and higher education sites—service sites for the target culture—interaction appears to be largely logical, branched and relatively passive, with the exception of the now ubiquitous invitations to email, comment, Facebook or Tweet. How many users take up the offer is difficult to assess. Information density is moderate to high, reflective of Hofstede's assessment of Argentine culture's moderate power distance index. Interestingly, with the exception of a moderately high uncertainty avoidance index, most of Hofstede's cultural dimensions have Argentine culture squarely in the middle. This is an example of why his 'masculine/feminine' dimension is so unsatisfying. The machismo so often associated with Latino culture doesn't align well with this midline masculinity reading, if one takes a gendered view. However, if one looks at the relatively high uncertainty avoidance score (86 where the high is 112 and the low is 8) there must be something at work here—the Argentines are so "in the middle" with regard to everything else in terms of attitudes and beliefs.

Looking at the types of images and the rhetoric of the verbal content, Argentines are more dramatic and emotional than the average. This is consistent with other Latin American cultures. They are also relational—meaning, a great deal of value is placed on relationship over data driven decisions. In most of the sites with multiple images roughly

Figure 10.9 (A) LA NACION LINE 2005

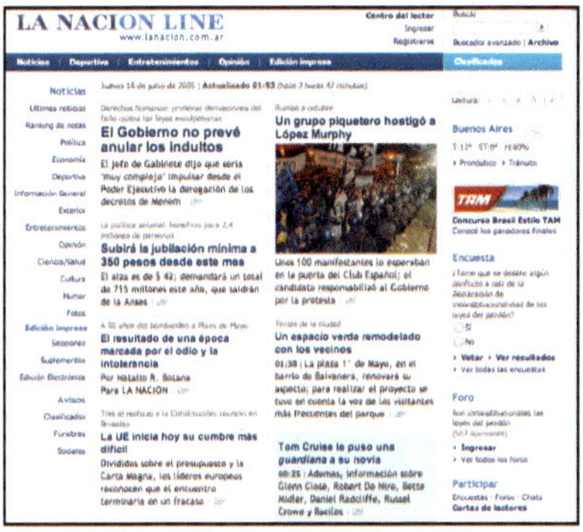

(B) lanacion.com 2013

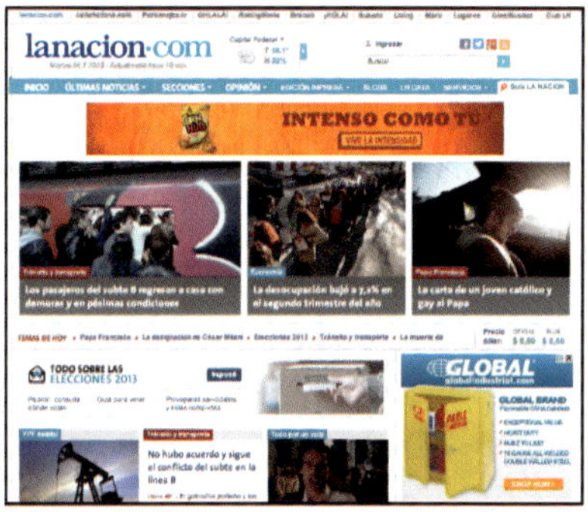

Figure 10.10 Cronica 2013 grid example

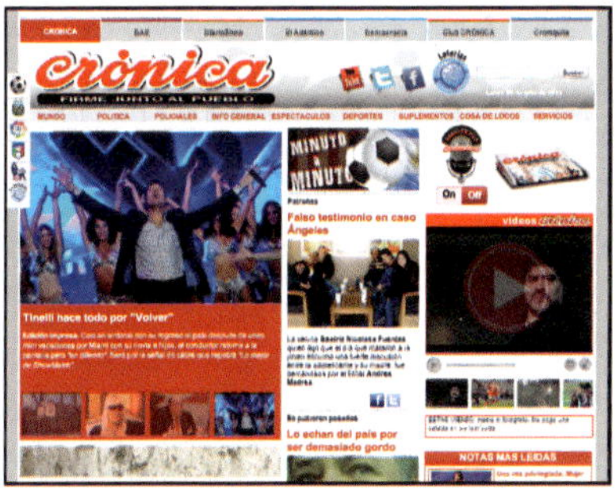

Figure 10.11 Creditcoop 2013

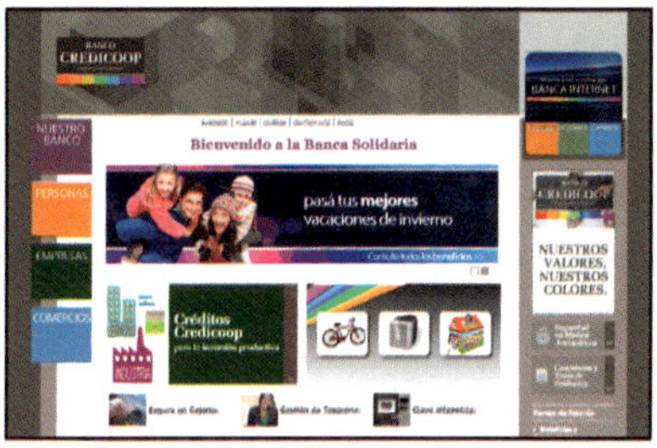

half of the images portray people in family or friend configurations, news sites being the notable exception. This is consistent with a polychronic culture where relationships are more important than tasks (Hall, 1983). Again, reinforcing a midline position, the value orientation in most of the websites appeared to be equally distributed between materialistic and quality of life focused. This could be a result of differences between the cosmopolitan urban center of Buenos Aires versus the general countryside. Urbanites identify closely with their European counterparts and place a great deal of value on formality and perceptions of status, ranking 40th out of 111 in *The Economist's* quality of life index (2005, p. 4). "Buenos Aires is one of the great fashion capitals of the world. In general, Argentines dress considerably more formally in business than their North American visitors. Considerable attention is also given to style and quality of material in dress" (Victor, 2013, n.p.). This formality combined with the passion of Argentina's national dance, the Tango, epitomize the scope of contrast that may have misled Hofstede to index the Argentines in the 'middle' of so many of his dimensions. What appears to be middle is just the warm you get adding hot and cool together.

Table 10.2 Cultural Aesthetic Grid - Argentina

Pattern Variable	http://www.cronica.com.ar/ (Figure 10.10) 2013 Tabloid newspaper based in Buenos Aires, Argentina
Color Palette Warm/cool Saturated/Muted National/Natural Contextual/Brand	Warm, vibrant, highly saturated colors: red and white Uses the Brand colors… same as in 2007, although the background in the masthead then was gray.
Content Modality (Image/Text) Ratio (% Visual, % Verbal) Density (High/Low)	Modality is 50/50 on this site above the fold. However, scanning down the page shows other sites show a slight preference for words over images. More than just headlines. 60/40 ratio. Moderate to low high density.
Information Flow Fast/Slow Poly-dimensional/Uni-dimensional	Information flow is fast - lots of descriptive images Polydimensional - readers can move to all parts of the website - drill downs and videos lots of ads
Time – mode Polychronic/Monochronic	Polychronic
Time – tempo Fast/Slow	Fast to download; lots of fast-paced animation including automated slide shows and advertising.
Audience Interaction Static/Dynamic Serial/Branched/Hypermedia Passive/Interactive	Dynamic interaction - typical of tabloid "yellow press" types of sites. Interactive - includes polls and comment sections. Ability to post to Facebook and Twitter
Information density High/Low	Moderate. Stories are tabloid length -- no real in-depth reporting. Confirms ranking in Hofstede (49/104)
Imagery Collectivist/Individualist (many people/one or two persons)	Individualist imagery -- people but mostly one or two. Celebrity driven so not too many multi-generational images
Context High/Low	High -- many of the subjects in news stories referred to by first name only - celebrity culture.
Attitude Rational/Emotional Objective/Relational	Highly emotional - very dramatic and colorful language Relational - traditional family ties are very important
Value Orientation Materialistic/Quality of Life	Mixed in this website - heavy advertising focus and yet, many of the ads were about life quality enhancing subjects
Usability Ambiguity tolerant/intolerant	Interesting - Home Page link called "Inicio"
Uncertainty Avoidance High/Low	Moderately high -- face saving related to machismo?

Example 3: Starbucks & Coca-Cola—Localizing "Web Flavors"

Figure 10.12 Comparison of Starbucks websites collected in 2007 and 2013.

(A) China 2007	(B) Germany 2007	(C) Japan 2007
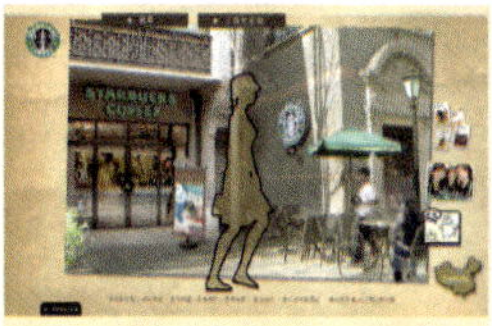	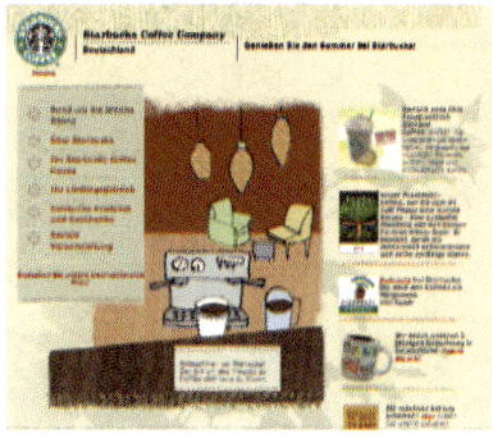	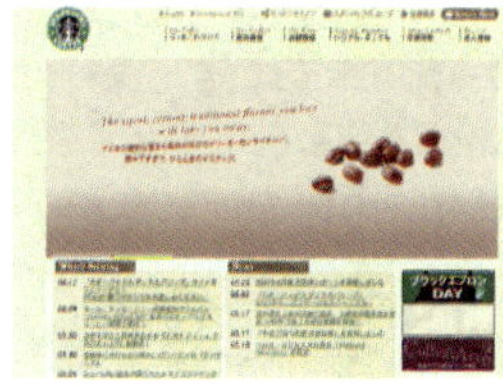
One person overlayed on the scene but it is a shadow charactrer, not a recognizable individual or celebrity.	No images of people at all. A great deal more verbal material. Organized, lots of information, but without much "white space."	Central image is a great example of Japanese *Kanzo*—spare with just the appropriate level of information.
(D) China 2013	(E) Germany 2013	(F) Japan 2013
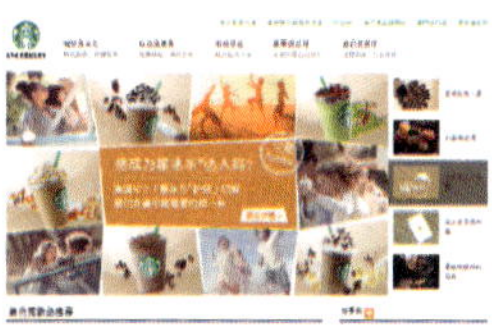	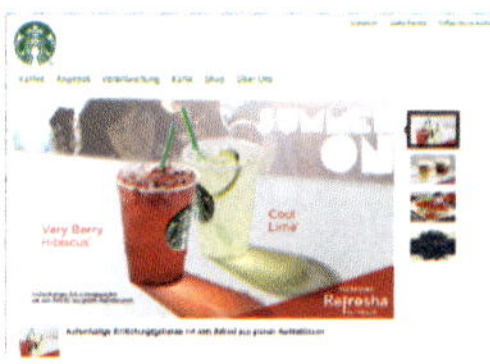	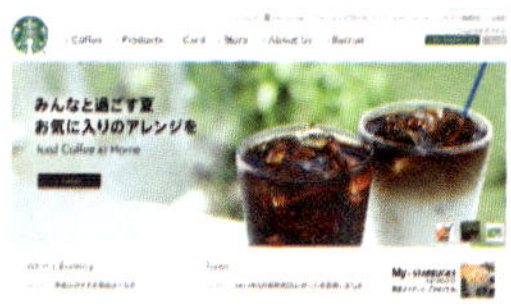
Fast-paced and dynamic central image (5 choices)	Static central image—click for alternatives (4 choices)	Static central image—click for alternatives (3 choices)

Analysis

The Starbucks websites, at the time that we captured the 2007 screens above (Figure 10.12), appeared to be localized—the sites looked like they were specifically created for that culture or country. For instance, the German site had a great deal more text than either the Chinese or Japanese site. The Japanese site had much more 'white space,' the central image conveying an almost Zen garden-like structure. However, all sites shared a brand aesthetic of neutral, natural colors.

In that same year, we noticed that Coca-Cola, another multinational beverage company, began using templates—a shift away from costly localization to the creation of four to six 'skins,' which are then used for various cultural groups based on some overall criteria. This strategy was largely employed by multinational websites to lower site development costs. With skins, one can use the same layout for cultures that appear to share a particular interest, for instance a soccer or football 'skin' to appeal to those cultures in which 'futbol' is a major activity. To see if these companies designers were doing their cultural aesthetic homework, we would have to look at the cultural patterns of these cultures and compare their beliefs, attitudes, and value orientations. The color palette would be of secondary concern because the brand colors would dominate as can be seen in Coca Cola's "Template 1" used for areas as diverse as Argentina, Egypt, and the Philippines (Figure 10.13).

Figure 10.13 Coca Cola's 2007 'Template 1' was used for the following countries/ cultures regardless of their cultural aesthetic similarities or differences:

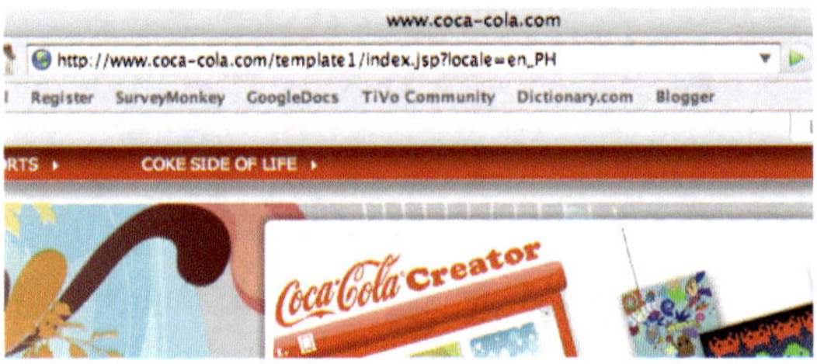

Argentina

Chile

Egypt

Philippines

United Arab Emirates

Recent (2013) observations of Coca Cola's websites indicate another shift in strategy. It is a combination of skins for the navigation wrapper with large video content areas for upscale technology markets and still images in the same area for low bandwidth cultures. A navigation wrapper usually provides a stable and static design for the common navigation. The navigational elements are pared down to the lowest common denominators. The content areas are then divided into "buckets" into which different types of content can flow, e.g., videos, still images, animations. Most of the videos are also available on YouTube indicating that there is a cross-media strategy for Coca Cola's advertising that combines the templating of website structure with the localization of television and video ads that can be easily slipped into the relevant content buckets.

This kind of strategy lets an enterprise repurpose its content to multiple platforms. Videos are cleverly done so that they can be dubbed into indigenous languages. The Spanish Coca Cola site featured a promotion (September 2013) called "Levantamos" (*stand up!*), a social awareness message targeted against sedentary, unhealthful lifestyles (http://www.youtube.com/watch?v=67PNtwM6Tso). In another video for the same campaign, Coca Cola reviews the statistics of obesity and its leading causes. The scenes in the videos are taken from all over the world, of people from many different cultures, capturing them in similar activities, essentially hybridizing and homogenizing the message. And there is an irony in the manufacturer of soft drinks, a leading cause of twenty-first century obesity, taking such a socially noble stand, as the many comments on the YouTube versions point out. This kind of 'glocalization'—videos that display culturally diverse images overdubbed with local scripts and languages—lead to an oversimplification of the message and of the issues they are meant to convey. They create a suggestion that the many and various problems experienced in these different parts of the world can all be addressed with a single solution.

The visual rhetoric embedded in this type of web design strategy—skins, wrappers and content buckets—is a rich area of study, but one that requires the researcher to do some form of screen capture that incorporates or traces the links to the video content. The one thing that having YouTube material embedded in the webpage does is yield some data on the number of views as well as provide viewer comments.

In a recent documentary, *Generation Like* (2014), Frank Koughan and Doug Rushkoff explores the impact that social media is having on all forms of media, but, in particular, the interconnected strategies of websites, and social media tools like YouTube, Facebook, Twitter, Instagram, Vine, Pinterest and whatever will be invented tomorrow or by the time this book gets into readers' hands. As users socialize the brands and causes they encounter on the web by clicking the LIKE button, they are also letting merchandizers know the kinds of media users respond to, where the user came from and where they go when they move off your website. The LIKE button becomes a portable market research tool. Users bring "Big Brother" along with them as they cruise around the Web looking for things to share with their social networks. This too is a rich area for social scientists observing all kinds of Web behavior.

Table 10.3 Cultural Aesthetic Grid - Starbucks Japan (2007)

Pattern Variable	http://www.starbucks.co.jp/
Color Palette Warm/cool Saturated/Muted National/Natural Contextual/Brand	Cool or neutral colors Muted shades Natural palette Brand driven
Content Modality (Image/Text) Ratio (% Visual, % Verbal) Density (High/Low)	Text modality <30% visual Very low density (lots of negative space - zen-like in its simplicity
Information Flow Fast/Slow Poly-dimensional/Uni-dimensional	Information flow is slow - images enigmatic, mostly words Sequential, unidimensional
Time – mode Polychronic/Monochronic	Appears to be monochronic
Time – tempo Fast/Slow	Slow
Audience Interaction Static/Dynamic Serial/Branched/Hypermedia Passive/Interactive	Nearly all serial text links to information that changes like news and events. Static corporate information links across the top - standard location. Passive interaction
Information density High/Low	In the areas where there was text, high density.
Imagery Collectivist/Individualist (many people/one or two persons)	Imagery did not feature people at all. Japan is traditionally considered a collectivist culture.
Context High/Low	Clearly high context. very little visual information to orient the user. Culturally embedded meaning.
Attitude Rational/Emotional Objective/Relational	Rational by default - no emotion or pathos in the page Neither objective nor relational
Value Orientation Materialistic/Quality of Life	In this screenshot, quality of life appears to dominate in the quiet contemplative imagery that takes up half the screen
Usability Ambiguity tolerant/intolerant	Absolutely no ambiguity - highly usable
Uncertainty Avoidance High/Low	Traditionally considered a high UA culture. This is apparent in this website. Clean flow.

Example 4: News Websites—"Print goes online"

Figure 10.14 Comparison of broadsheet news websites collected in 2007 and 2013.

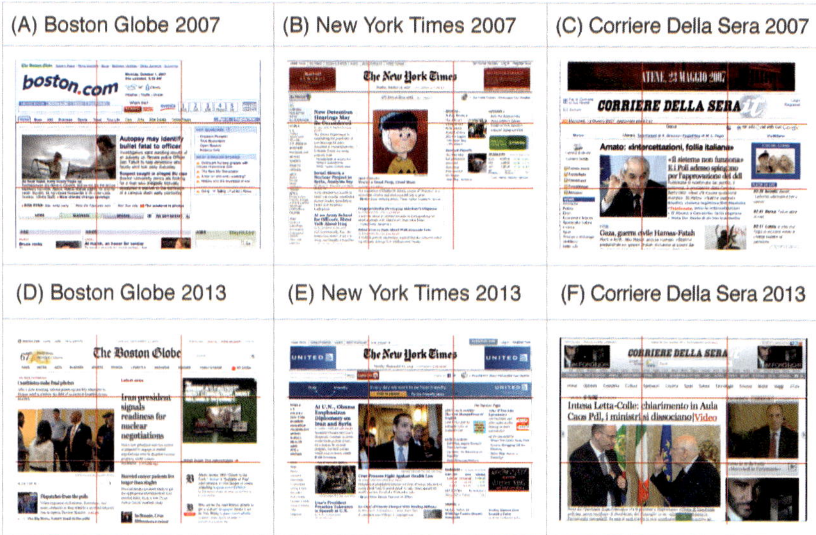

Analysis

In this example, we are comparing three well-known broadsheet newspaper websites. By "broadsheet" we mean the traditional newspaper format where the final fold is horizontal so that only half the front page news stories are visible at the newsstand. The other format for daily news in print is "tabloid," e.g., Argentina's *Cronica* (analyzed in our second example in this chapter) or News Corp's *The Sun*.

The print broadsheet newspaper has given us much of the pattern language we use for information websites. "Masthead" and "above (or below) the fold," when we mean what is visible in the browser window at first sight, are two such examples. Often referred to as "serious" newspapers, broadsheets had a difficult transition to being online at first. An effective business model for delivering news on the web had not yet been formed back in 2007. Some would argue that a sustainable business model still hasn't been found, partly because rapid shifts in technology make it hard to keep up and zero in on what newspapers can do to make a profit online. However, staking out territory on the web in the early to mid-2000s was critical if one didn't want to be left behind one's competitors. In the first forays into online news publishing, some long-lived and respected brands over-steered themselves into dot-coms that obscured their identity as major household names. "One approach [to getting on the Web] focused on trying to convert the print brand to a dot-com by changing the logo and brand name to a URL address." This was true of the Atlanta Journal Constitution (ajc.com) and the Los Angeles Times (latimes.com). "Many of these news sites not only changed their name but perhaps went into partnership with other content providers, subjugating their brand to another authority" (O'Connell, 2004, p. 6). The *Boston Globe* did this when they morphed to boston.com, as did *The Philadelphia Inquirer* and *The Philadelphia Daily News*, two separate papers that both appeared under the URL philly.com. However, other broadside newspapers leveraged their brand, not only keeping their names intact, but also using their distinctive type set in the masthead.

In this case study, we look at three broadside newspaper examples, the *Boston Globe*, a large establish regional American paper, the *New York Times*, also a large established regional paper with an international reputation and readership, and *Corriere Della Sera* (The Evening Courier), one of Italy's national newspapers with editions for all of Italy's major cities.

In addition to comparing these online news publishing sites to each other, we also make a longitudinal comparison with screenshots taken in 2007 and, again, in 2013. These comparisons establish two things: 1) that the overall structure of a newspaper online website closely mirrors its print counterpart and 2) that despite the similarities of broadsheet or tabloid newspapers around the world, and subsequently their mirrored websites, in terms of structure, there are still cultural aesthetic differences that can be identified using our pattern language.

Of the many patterns outlined in our grid, the patterns that concern us most with this genre of website deal with how content is displayed. Just as with their print editions, very little of a journalism website's content will concern itself with advertising except in the animated banners at the top or the right-side margin where advertising traditionally resides. Because of the longstanding divide between advertising and editorial, a reputable newspaper website will have clearly defined areas for advertising and for editorial or news. This does make it easier for the researcher/observer to grid out the content areas and establish which modality is preferred by the site's target audience. It is more difficult when we go down a level or two to the individual stories where we are likely to find interstitial blocks of advertising in the middle of the news story. This appears to be an idea borrowed from television advertising when we are periodically served with commercials.

One of the major shifts for traditional newspaper publishers upon going online was the use of color images and video. These were options either technically not possible (video on paper) or economically unrealistic (pre-laser 4-color separations for color images). Newspapers that would have used few, if any, color images were now able to have all images be in color—indeed, it would have been more work, not to mention slightly odd, for photos to be grayscale online as they would be in the print edition. Even the *Wall Street Journal* has moved away from their signature hedcuts, small pencil-dot portraits, for their online offering.

In all six of our examples in Figure 10.14, large color images are used for the headline story and smaller color photos are used as entry points for other stories. In the two American sites (*Boston Globe* and *New York Times*), the relative sizes of the images for both lead stories and secondary stories remained the same over time. However, for *Corriere Della Sera*, the visual verbal ratio for content jumped significantly from 2007 to 2013. Indeed, in the 2013 screenshot of the home page, the only verbal content is the main headline and the captions for the two large content images. The other large image is advertising for Tom Ford Noir.

Figure 10.15 Interactive areas of newspaper websites 2013.

(A) Boston Globe

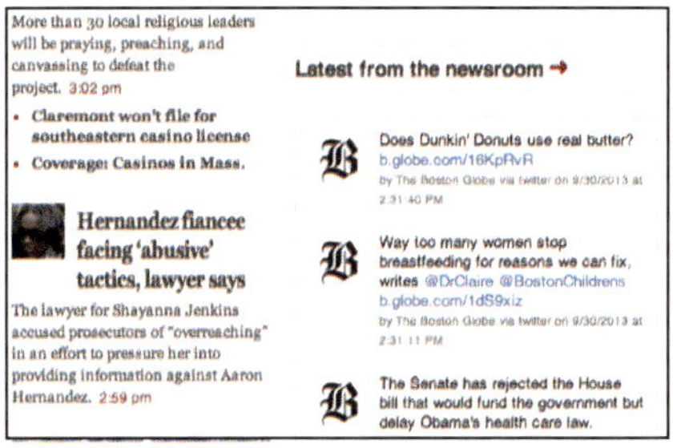

(B) New York Times

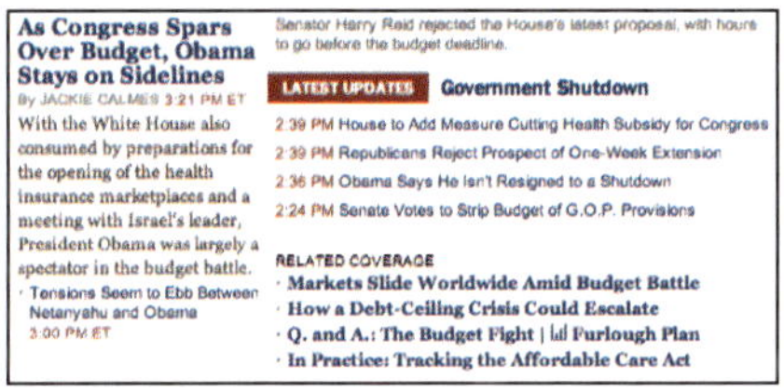

(C) Corriere Della Sera

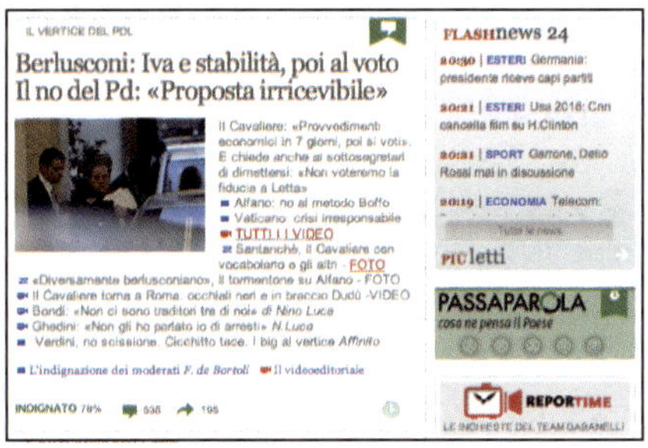

This difference suggests that the Italian news site is exhibiting a cultural preference for visual modality. In addition, the Italian site has a higher level of interactivity. The "Flash News 24" area, placed well "below the fold," features a scrolling list of breaking news stories (Figure 10.15). Small icons placed under and to the side of the story's image lead to more interactive elements including editorial video from CorriereTV.

In the same areas on both the *Boston Globe* and *New York Times* sites, the lists were static, i.e., not moving. However, on the *Boston Globe* site, the list could be manipulated using the mouse. Of the three breaking news areas, the *New York Times* was by far the denser with little white space and a small font size. The *Boston Globe* had more white space, largely because it uses a blackletter "B" icon next to each breaking story link.

An interesting comparison on the structure of the newspaper site genre is where the navigation is located. Most newspapers have several "departments" or "sections" that might include national news, world news, politics, sports, entertainment, fashion, health, opinion, etc. The *Globe* and *Correire* locate their content navigation horizontally across the top just under the masthead. The *New York Times* locates theirs along the left side, breaking the content links into major headings and sub-categories that include items for which they have a reputation, like the crossword puzzle—a cultural icon with its on documentary, "Wordplay" (2006) about the NYTimes crossword editor, Will Shortz, and his work. The New York Times navigation, in contrast to the two other examples, appears deeper, richer, promising more content, suggesting that their readers, not surprisingly, have a strong preference for the verbal mode and dense, deep levels of information.

In the early days of the web, newspaper websites, as with most other websites, suffered from relatively poor usability. In a 2004 usability review of ten regional American newspapers, I made the following observations: "The most egregious of these design decisions is the removal of the underlining of links in a mistaken attempt to 'clean up' the look of a webpage. The result is that users conduct a scavenger hunt with the mouse, passing over text to see if the cursor can identify hot-linked content" (p. 12). The perpetrators were the *Los Angeles Times*, the *Philadelphia Inquirer*, and the *Arizona Republic*. The *New York Times* was reported the "best in class" for its affordance in that study, with the *Houston Chronicle* as a runner up. In 2013, the links on the *New York Times* site are still colored, but one must hover over the link to see the underlining that confirms that the text is a link. Of course, over the years, readers have become accustomed to headlines being linked so it is the website that *doesn't* link its headlines that would lose the usability test now. However, having the link change color once it has been visited is still not universally done in newspaper websites—places where the reader could use a little help to know what they have clicked and read and what is still there to explore.

The use of columns in newspaper websites to organize content is another pattern carried over from the print version. In the very early days of the web, the layout capabilities of HTML did not allow for the sophisticated use of columns. Some newspapers basically created an image of their paper version and posted that. Many newspapers still do that because a newspaper is a "record" document. The newspaper "of record" is where legal notices would be published to ensure that they have been publicly posted. With the advent of the Web, this function has been largely shifting to online sources, although one might argue that universal access to the Web is still a dream for the future. Paperofrecord.com provides a searchable archive of printed newspapers, particularly historical ones that might not otherwise be available except in the basements of local libraries.

Layout designers have established, and tested, the various combinations of columns and column sizes for ease of readability. Newspaper layouts tend to be divided into three, four, five, or seven column multiples, the odd numbered column layouts having more balance than those that are even numbered. Some content areas take up the space of two columns, as with the *Wall Street Journal* example in Figure 10.16. This would be described as a five column grid with the first content area 1 over 1, and the next two content areas 1 over 2 (columns). If one divides up the visible space using the narrowest column as a guide, it is not difficult to determine how the space is divided and how the content areas are gridded.

Figure 10.16 Use of still color images and video with interactive charts in today's online news sites. Interactivity is indicated with the "play video" arrows. The stockmarket table updates periodically throughout the day.

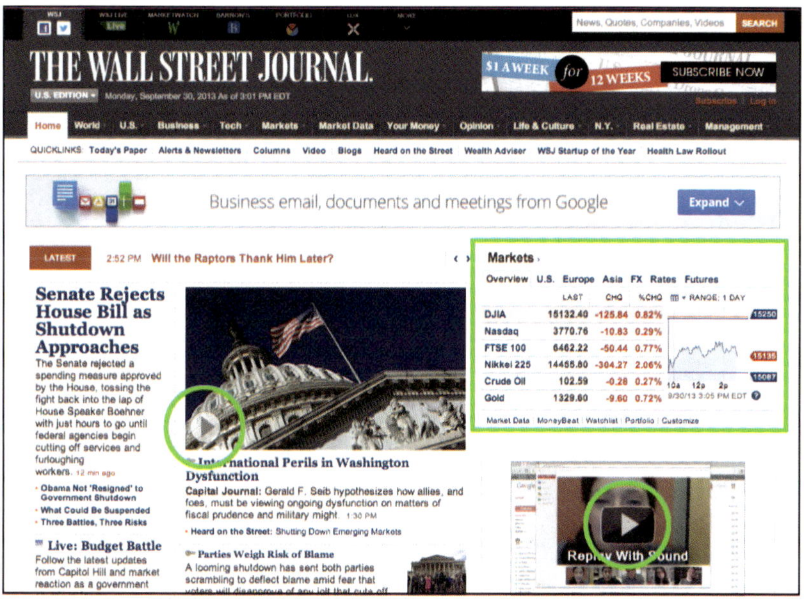

In comparing the American newspaper sites to *Correire Della Sera*, one can observe again the dominance of the verbal mode in the layout of the *Globe* and the *New York Times*, versus the visual mode of the Italian site. Where the American sites use a five or seven column grid, *Correire* now uses a three column grid with the main content area 1 over 2—a huge block of image dominating the screen.

This example provided a way to look at the structure of a website pattern based on a functional genre—in this case newspaper websites. Websites that are companions to other news media, television, radio or magazines, will exhibit similarly inherited characteristics derived from the original medium. Despite the structural commonalities that can be found across all newspaper websites, there are still cultural aesthetic differences that can be observed by noting the visual/verbal modality and ratio and the density of the layout. The commonalities are bound in the original medium's reader/viewer culture and determined largely by the technology of the original medium itself.

Table 10.4 Cultural Aesthetic Grid - News Italy *Corriere Della Sera* (2013)

Pattern Variable	http://www.corriere.it/
Color Palette Warm/cool Saturated/Muted National/Natural Contextual/Brand	Cool or neutral colors Muted shades Natural palette Brand driven
Content Modality (Image/Text) Ratio (% Visual, % Verbal) Density (High/Low)	Images dominate the page. The visual/verbal ratio is clearly 90/10 with very high density on the page. No white space at all.
Information Flow Fast/Slow Poly-dimensional/Uni-dimensional	Information flow is quite high as it is nearly all coded in images. It also appears to be poly-dimensional with few if any restrictions.
Time – mode Polychronic/Monochronic	Polychronic, providing many entry points and activities for users to explore.
Time – tempo Fast/Slow	Not observable with this kind of website.
Audience Interaction Static/Dynamic Serial/Branched/Hypermedia Passive/Interactive	Audience interaction was dynamic, with the links both serial and branched. Interactive.
Information density High/Low	Information density, on the surface, appeared low. But there are many points of entry and many modes in which the information was presented.
Imagery Collectivist/Individualist (many people/one or two persons)	Images were largely photojournalistic in nature and of individual politicians or celebrities.
Context High/Low	Low context over all
Attitude Rational/Emotional Objective/Relational	Strong use of images and color suggest an emotional attitude with relational overtones. Strong sense of community/nationalism.
Value Orientation Materialistic/Quality of Life	The advertising suggests a materialistic value orientation, the information content did not suggest either one.
Usability Ambiguity tolerant/intolerant	Usability was good.
Uncertainty Avoidance High/Low	Low UA

Afterword

Moving Forward

> *Cultural objects including images, languages, and hairstyles now move ever more swiftly across regional and national boundaries. This acceleration is a consequence of the speed and spread of the internet and the simultaneous, comparative growth in travel, cross-cultural media and global advertisement.*
>
> —Arjun Appadurai, "How Histories Make Geographies"

If there is anything the past two decades of the Web have shown us, it is that the Web is something that never stands still. It is, every second of every day, moving forward into unknown, uncharted technological territory. Sometimes that is defined by platforms and tools we know and sometimes, as with the introduction of the iPad in 2010, our relationship to the Web and its content changes dramatically. We can only look at, identify, and study what we have seen in the past and what we are seeing now.

And we can only create content for the Web using the tools we have now. The tools that will be developed later today, tomorrow or in months and years to come, will continue to push content creators along the technology/design cycle. It will never settle down or become static as long as there are resources to create, update and redesign, and a public who expects the content on the Web to match the power and mobility of the digital appliances they have in their hands. Technology is a driver of design. It may also be a driver of cultural aesthetics.

In the beginning, there was HTML, the formatting or pattern language of the Web (Berners-Lee, 1993). However, HTML 1.0 was severely limited in its ability to do much of what we think of as formatting at all. Like Ford's Model T, you could have any kind of website you wanted as long as the background was gray (default setting), the text was wall-to-wall, and users didn't mind waiting for 5 minutes to download a poorly rendered, pixelated image limited to 256 colors. Today, there is a vast library of software that helps people use the Web's pattern language (HTML 5.0) to produce the aesthetic result they want—even in languages with cyrillic, arabic and glyphic typesets. In the brief two decades of the Web, we have come a very long way. The Web is the most global entity in the world. It is arguably the fastest growing communication and economic engine of twenty-first century globalization.

The concept of *globalization* has long been the purview of political science and economics but it is also the concern of anthropology, sociology, communication and cultural studies. Defining globalization quickly becomes an tangled exercise:

> Economic, political, cultural, and social dynamics are not simply different facets of a single globalization; rather they are each prisms through which globalization takes shape and is experienced and mapped differently, yet they all mingle and interpenetrate as well. Thus, as each of the social sciences holds a different perspective on globalization, even a multidisciplinary approach to globalization still resembles eating soup with a fork. (Nederveen Pieterse, 2009, p. 14)

Perhaps it is less important to define globalization than it is to recognize its presence. If we are concerned with how a culture is mediated on the Web, we must turn our attention to the how the Web impacts, and is in turn impacted by, globalization.

While the Web is a medium, its content is media made global—words, images, sound, video, everywhere and on every digital device. Mark Deuze writes, "You live *in* media. Who you are, what you do, and what all of this means to you does not exist outside of media." He goes on to paraphrase Marshall McLuhan, "Media are to us as water is to fish. This does not mean life is determined by media—it just suggests that whether we like it or not, every aspect of our lives takes place in media" (2012, p. x). The Web has become the major access point for *media as environment.* If we accept the argument Deuze makes when he writes, "[T]oday, the uses and appropriations of media can be seen as fused with *everything* people do, *everywhere* they are, *everyone* people aspire to be," we are acknowledging the impact of globalized media.

But the Web is also impacted by the global cultural forms and flows that move, not only information, but all manner of cultural objects, from images to hair and clothing styles, at evermore accelerated speeds via, not only the Internet, but also travel, migration and the power of global advertising and cross-cultural media (Appadurai, 2010). Within these flows, the new global cultural economy is a "complex, overlapping, disjunctive order that cannot any longer be understood in terms of existing center-periphery models... nor is it susceptible to simple models of push and pull,... surpluses and deficits,... or consumers and producers" (1996, p. 32). Appadurai proposes, in *Modernity at Large*, that we look at these global cultural flows as a web of relationships between a collection of '-scapes': ethnoscapes, mediascapes, technoscapes, financescapes and ideoscapes.

> These landscapes thus are the building blocks of what (extending Benedict Anderson) I would like to call *imagined worlds*, that is, the multiple worlds that are constituted by the historically situated imaginations of persons and groups spread around the globe. (1996, p. 33)

Image, images, imagined worlds—these are the things in motion that circulate through cultural flows and networks of connectivity leveraged by the Web. More importantly, for our investigations, these two concepts, the "work of the imagination" and the "circulation of forms" lead to a negotiated, multi-sited, less linear local product with its own aesthetic constituted of multiple points of origination.

Perhaps, for the cultural pattern-seeker, this becomes the most challenging effect of globalization on the Web—the *mélange* of the cultural and technical that has no specific point of origination. We see this when cultural forms, largely, but not exclusively, originating in the West, are adopted and adapted to the conditions of local communities in a type of *glocalization*. Like Afropop, a musical style that draws on many ethnic styles, these cultural products are the result of a type of fusion that makes it difficult to see where one cultural influence picks up and another leaves off simply because the actual cultural result resembles neither. It becomes its own thing, semi-rooted, belonging to many cultures and to none.

Of course, the variable that is difficult to pin down, is the effect globalization has on a particular audience's aesthetic preferences. The question for those of us looking at the Web using our pattern language is whether the effect of globalization on the aesthetics of a culture is, like the technology/design cycle, a loop or wave whereby there is an appearance of increased homogenization followed by a resurgence of diverse cultural expression. Or, is the effect a more permanent progression from cultural diversity to aesthetic uniformity? And how do we track the subtle shifts from one aesthetic expression to another? These observations are further complicated by underlying movements in politics and economics—themes of "ethnic politics (ethnic cleansing and new nationalisms) and religious fundamentalism, suggesting a link between globalization and local identity politics and a combination of integration and fragmentation" (Nederveen Pieterse, 2009, p. 1). Anyone trying to decode a website's cultural aesthetics will have to have more than surface knowledge of what is happening on the ground as well as on the Web.

Digital communication technology makes it possible for just-in-time, lot-size-of-one, production to take place wherever in the world the economics make financial sense. This, in turn, challenges the idea of national boundaries and their role in regulating and maintaining the integrity of the Web. In some cases, this sets one super power against another—as with the People's Republic of China, arguably the world's most populated country, blocking Google™, arguably the world's largest search engine (Huffington Post, 2011). The Chinese government's desire to censor search results was in stark violation of Google's mission "to organize the world's information and make it universally accessible and useful" (google.com). In other cases, national boundaries are digitally permeable or exist in name only. Obtaining a country's domain extension may, or may not be strictly regulated. Multinational enterprises have several country extensions allowing them to appear as digital citizens of multiple nations. Using national domain extensions as a means for identifying cultural aesthetics, then, is faulty logic. Unfortunately, there are no actual *cultural* domain extensions.

For cultural purposes, it is perhaps best for us to see nation states, and their shifting borders, as a "*grid* that has been temporarily superimposed upon a deeper and ongoing stratum of human migrations and diasporas" (Nederveen Pieterse, 2009, p. 36). The culture is not the nation, and the nation is not the culture. Indeed, because of the periodic shifting of geo-political borders and diasporal migration, any country can have within it many cultures, and a particular culture can have pockets all over the globe. If we really want to identify and explore the cultural aesthetics of a community, we have to find their *neighborhood*, "the actually existing social forms in which locality, as a dimension or value, is variably realized. Neighborhoods, in this usage, are situated communities characterized by their actuality, whether spatial or virtual" (Appadurai, 1996, p. 179). A culture's Web neighborhood is the virtual milieu discoverable by a careful analysis of the visual virtual artifacts: websites, films, print ads, because all are constructed self-representations of that culture's identity.

The Not-so-final Word

Throughout this book I have attempted to outline how to watch and track a moving target. For the student of human cultural expression, there is nothing quite as plastic a canvas as the World Wide Web. It is always growing. While there are still parts of the world that have no virtual real estate there, each day sees more and more parts of the world show up. It is an index of the "haves," a locus of the economic divide, and yet, conversely, it is an entrepreneurial equalizer. As a popular *New Yorker* cartoon reveals, "On the Internet, nobody knows you're a dog" (Steiner, 1993). Small start-ups with a keen idea of the aesthetic preferences of their target audience can, at relatively low cost, look and behave online like one of their bigger competitors. Global agoras like Etsy.com help craftspeople and artisans from all over the world showcase their wares to a worldwide marketplace. Sites like Pinterest allow people to "pin" images and videos they want to share with others on "pinboards"—the global equivalent of a cork board with endless applications like wedding planning, travel planning, cooking, home decor.

At the same time as the Web is growing, it is morphing. It is increasingly difficult to determine points of origin as certain top-level domain extensions (TLDs) have become the hallmark of clever marketing ploys known as domain hacks. Notable examples include del.icio.us (United States), fold.it (Italy), instagr.am (Armenia), and bit.ly (Libya). These are minor disjunctures, twists in the crossroads of the media/technoscape for the cultural aesthetic pattern-seeker. If the ubiquitous ".com" is the generic extreme for corporate identities, ".me" (Montenegro) is becoming a popular extension for the personal website. As mobile devices become the appliances of preference for accessing internet content, cultural aesthetics will have to change, yet again, to accommodate smaller screen sizes and interfaces.

These are exciting, if sometimes confusing and frustrating, times. However, the one thing that doesn't change very quickly is the human at the beginning and end of it all. We may alter how we read from paper to computer screen to tiny screen, but our eyes still function the way they have for millennia. We still see the same colors and shapes. We still attach cultural meaning to the symbols we see and use to communicate. And while we are not likely to change what we call a certain color, we quickly adapt to changes in how we use certain shapes. A few decades ago a colon and a parenthetical mark were simply punctuation. Today, they combine to form a universally accepted and understood emoticon :). It is this very ingenuity and flexibility that we rely on when we go exploring, seeking the patterns that tell us who we are. If we refine our pattern-seeking sensibilities and can identify the pattern of a particular cultural neighborhood, we can better communicate with the folks who live there.

Glossary of Pattern Language Terms

affect - emotional (pathos) ~ part of a value pair that is a subset of Hofstede's masculinity/femininity cultural attitude pair. People and cultures that show a preference for drama, exuberance, or emotional appeals can be characterized as valuing the emotional.

affect - rational (logos) ~ part of a value pair that is a subset of Hofstede's masculinity/femininity cultural attitude pair. People and cultures that show a preference for facts or fact-based arguments can be characterized as valuing the rational.

ambiguity ~ a word, expression or image that can be interpreted or understood in more than one possible way.

attitude ~ "a relatively enduring organization of beliefs around an object or situation predisposing one to respond in some preferential manner" (from Rokeach)

attitude - collectivist ~ a relational orientation that is based either on family or work/community groupings and driven by the expectations of others, particularly those of community and those in authority

attitude - individualist ~ a relational orientation driven by internal motivations where people tend to view themselves as individuals, see success as a result of individual effort, and tend to emphasize the needs of individuals over the collective.

audience interaction ~ the extent to which a user interacts with a website by manipulating elements or contributing content. Minimal to no interaction is considered "passive." Lots of interaction is considered "active."

brightness/value ~ measures how light (tint) or dark (shade) the color is or its value.

cognitive psychology ~ branch of psychology that deals with mental processes: how people think, how they take in sensory information, perceive its significance, and attach meaning.

cognitive thread ~ the sequence of concepts that are held actively in visual and verbal working memories, together with the links between them (Ware, 2008, p. 138)

collectivist/individualist ~ two contrasting cultural orientations. See "attitude"

color palette ~ the collection and range of colors used in a particular context based on hue, saturation, brightness and temperature.

connotative meaning ~ culturally situated meaning of a word, phrase or image

content (constant and dynamic) ~ refers to the kind of content information one finds on a website. Constant content does not change and is usually structural (About Us, Contact Us), decorative (tag lines) or navigational (menus). Dynamic content is generally ephemeral and can change from day to day or even minute to minute (front page news stories, weather conditions, interest or exchange rates). Content such as blog posts display both conditions—they are always findable, however they only occupy the "above the fold" position on the home page until the next new post is posted.

content density ~ the amount of white or negative space on a page or webpage. High content density means there is little negative space and the content is closely packed. Low content density means there is a good deal of negative space dividing up the content areas.

context (high/low) ~ describes how people in a culture perceive and communicate meaning and which form of communication, or coding/decoding, is valued by a culture. There are high-context situations and pockets within every culture, no matter how low-context that culture is in general.

cultural aesthetic ~ aesthetics in websites that are culturally derived and to which we expect audiences from those cultures to respond.

cultural identity ~ (simple) a person's sense of their own culture; (complex) "one, shared culture, a sort of collective 'one true self', hiding inside the many other, more superficial or artificially imposed 'selves', which people with a shared history and ancestry hold in common..." but also "a matter of 'becoming' as well as of 'being'. It belongs to the future as much as to the past" (Hall, 1993/1994, p. 393).

decoration ~ visual material that does not serve a navigational or informational purpose on the website. Decorative material may serve the purpose of branding as in a logo.

deixis ~ from the Greek "to show", words that are used as spatial references particularly in web buttons and links.

denotative meaning ~ the literal, factual meaning of a word, phrase or image

dynamic ~ having a characteristic of constant change; moving, active

emphasis ~ in typography and images, emphasis is when an element stands out from the other elements around it. In typography, emphasis can be conveyed using italics or bolding. In images, emphasis can be created by heavy borders or the use of a "drop shadow" suggesting dimensionality. The use of color, especially yellow, can be used to create emphasis.

ethnoscape ~ "the landscape of persons who constitute the shifting world in which we live: tourists, immigrants, refugees, exiles, guest workers, and other moving groups and individuals constitute an essential feature of the world and appear to affect the politics of (and between) nations to a hitherto unprecedented degree" (Appadurai, 1996, p. 33).

flow ~ (work) the perception of moving from one task to another; (information) the perception of drilling down into deeper, richer content or the perception of access to information based on authority or privilege, such as a paid subscriber vs. an unpaid visitor to a site.

framing ~ refers to the control of the information and therefore is equally connected to information flow, or what Hofstede called "power distance."

Gestalt (principles) ~ a set of natural rules that explain why we perceive certain objects to be grouped or not grouped. Gestalt is a German word for a pattern or the essence of a form.

globalization ~ a process that encompasses the causes, course, and consequences of transnational and transcultural integration of human and non-human activities. (Al-Rodhan & Stoudmann, 2006)

high context ~ require a high degree of homogeneity in traditions, rituals and practices; focus on in-group behaviors, tend to be more invested in face saving than low-context cultures

hue ~ the primary attribute of a color; it is the property that makes the color discernible as red or green or blue; pure color as determined by wave-length.

imagery ~ the likenesses of things in pictorial form; the mental visual representation of things; the vivid description of things.

information (content) ~ content that provides knowledge, data

information flow ~ the direction in which information is communicated, e.g., hierarchically (see uni-dimensional flow), laterally or organically (see poly-dimensional flow)

Information processing ~ in cognitive psychology, the manner in which sensory and arbitrary data are gathered, manipulated, imbued with meaning, stored, retrieved and categorized. Can also be used for similar functions in computer and internet engineering.

information/task complexity ~ the amount and type of information needed to operate confidently in a task domain

intensity (color) ~ the strength and purity of a color. Also referred to as a color's "saturation".

interactive ~ having the capacity to involve the user in a two-way activity.

juxtaposition ~ when two objects, images or ideas are combined together such that the differences between them create a new message, sometimes discordant, can be humorous.

lifestyle-materialistic ~ the value orientation that wealth, worldly possessions and physical well-being represent the "good life."

lifestyle-quality of life ~ the value orientation that a balance of work, family, health and time to enjoy these constitute the "good life."

linearity ~ having one dimension, one direction.

low context ~ tend to be explicit and verbal; have many casual relationships or acquaintances.

materialistic/quality of life ~ see "lifestyle"

mediascape ~ "refer to both the distribution of the electronic capabilities to produce and disseminate information (newspapers, magazines, television stations, and film-production studios, [and social media])... and to the images of the world created by these media" (Appadurai, 1996, p. 35).

metaphor ~ a symbol, the use of a word, object or image to represent something else.

mirror neurons ~ brain cells that mediate the behavior of others and are important to our ability to recognize the behaviors or language of others and imitate them.

modality ~ the dominant mode that content or information falls into, e.g., verbal mode (words), symbolic mode (mathematics) or image mode (pictures). Can also refer to interaction modes of passive or interactive.

narration ~ the story that is being told, including the voice used to tell the story.

navigation ~ the process or act of knowing where one is and finding one's way to another location. On the web, URLs and links provide the tools for navigation.

objective/relational ~ see "orientation"

orientation - objective ~ an orientation toward decision making that relies heavily on data driven reasons for making a choice or purchase.

orientation - relational ~ an orientation toward decision making that relies heavily on the opinions of others in one's relational groupings. Viral marketing operates on exploiting a relational decision orientation.

pattern language ~ the terms that describe a collection of multi-use, interdependent patterns that act as a shorthand when communicating what we are seeing or building.

perception ~ the awareness of the elements in our environment based on sensory input. These sensory perceptions are translated into coherent patterns of meaning in the brain.

point of view ~ the perspective from which one considers what one sees or considers.

poly-dimensional flow ~ organic and lateral flow of information, bottom-up driven and moving unrestricted through the websphere.

rational/emotional ~ see "affect"

saturation ~ how intense the color is; relatively measured as a range between a color's highest saturation and its least, or gray, where the closer to gray it gets the more muted the color is.

semiotics ~ a general theory of signs and symbols that deals with communicative behavior and how these construct our understanding of what we see.

stability ~ the state or quality of being stable, reliable

static ~ unchanging

symbolism ~ using symbols to represent ideas or things; attributing a symbolic meaning to a color, shape or some combination of these

technoscape ~ "the global configuration... of technology and the fact that technology, both high and low, both mechanical and informational, now moves at high speeds across various kinds of previously impervious boundaries" (Appadurai, 1996, p. 34).

temperature (color) ~ the perceived base color of a color palette. Warm colors have a yellow base. Cool colors have a blue base. There are warm blues and cool reds.

time mode ~ relates to how a culture perceives and manages time. At one end of the spectrum is the preference for punctuality, schedules, and the need to tightly manage those schedules. At the other end is a preference for placing the importance of the task at hand and personal and professional relationships first, putting attention to the schedule a distant second.

time tempo ~ refers to the speed or pace at which someone within a culture accomplishes everyday tasks.

uncertainty avoidance ~ a desire to avoid the discomfort of ambiguity or uncertainty or the possibility of making errors

uni-dimensional flow ~ hierarchical and access, top-down and controlled sharing.

usability ~ a quality attribute that assesses how easy user interfaces are to use. The word "usability" also refers to methods for improving ease-of-use during the design process (Jakob Nielsen, http://www.nngroup.com/)

value orientation ~ a social construct that describes a person or culture's worldview, principles of right and wrong, and value system.

visual/verbal ratio ~ an indicator of a cultural preference for believing what one reads (words) or what one sees (images). (Kress and Van Leeuwen, 1996)

References

Albers, J. (1968) Oral history interview with Josef Albers, 1968 June 22–July 5, Archives of American Art, Smithsonian Institution. Interview with Sevim Fesci for the Archives of American Art.

Albers, J. (1971). *Interaction of color.* Revised and expanded edition. New Haven, CT: Yale University Press.

Alexander, C., Ishikawa, S and Silverstein, M. (1977). *A pattern language: Towns, buildings, construction.* New York: Oxford University Press.

Al-Rodhan, N. and Stoudmann, G. (2006). Definitions of globalization: A comprehensive overview and a proposed definition. *Program on the Geopolitical Implications of Globalization and Transnational Security.*

Appadurai, A. (1996). *Modernity at large: Cultural dimensions of globalization.* Minneapolis: University of Minnesota Press.

Appadurai, A. (Ed.). (2001). *Globalization.* Durham: Duke University Press.

Appadurai, A. (2010). How histories make geographies: Circulation and context in a global perspective. *Transcultural Studies, 0*(1), 4–13. doi:10.11588/ts.2010.1.6129

Appadurai, A. (2012). Disjuncture and difference in the global cultural economy. In F. Lechner and J. Boli (Eds.) *The globalization reader.* Malden, MA: John Wiley & Sons. pp. 95–103.

Ball, M. (1998). The visual availability of culture. In J. Prosser (ed.) *Image-based research*, 131–146. London: Routledge.

Barber, W. and Badre, A. (1998). Culturability: The merging of culture and usability. Proceedings of the 4th Conference on Human Factors & the Web. June 5, 1998, Basking Ridge, NJ. Retrieved from http://zing.ncsl.nist.gov/hfweb/att4/proceedings/barber/

Barthes, R. (1957). *Mythologies*. Paris: Seuil.

Barthes, R. (1972). *The rustle of language*. Trans. Richard Howard. Berkeley: University of California Press.

Barthes, R. (1977). *Image, music, text*. New York: Hill and Wang.

Baxandall, M. (1988). *Painting and experience in fifteenth century Italy*. London: Oxford University Press.

Berger, J. (1972). *Ways of seeing*. London: Penguin Books.

Berlin, B. and Kay, P. (1999). *Basic color terms: Their universality and evolution*. Stanford, CA: CSLI Publications.

Berners-Lee, T. and Connolly, D. (1993). Hypertext markup language (HTML): A representation of textual information and metainformation for retrieval and interchange. Internet Draft, IIIR Working Group. Retrieved at http://www.w3.org/MarkUp/draft-ietf-iiir-html-01.txt

Bolter, J. D. and Gromala, D. (2003). *Windows and mirrors: Interactional design, digital art, and the myth of transparency*. Cambridge, MA: MIT Press.

Bowers, J. (2008). *Introduction to two-dimensional design: Form and function*. 2nd Edition. New York: John Wiley & Sons.

Brown, B. (1977) Face saving and face restoration in negotiation. In D. Druckman (Ed.), *Negotiations: Social-Psychological Perspectives*. Sage. pp. 275–300.

Bryan, M., Cameron, J. and Allen, C. A. (1999). *The artist's way at work*. New York: William Morrow and Company.

Buchanan, R. and Margolin, V., Eds. (1995). *Discovering design: Epxlorations in design studies*. Chicago: The University of Chicago Press.

Burmark, L. (2002). *Visual literacy: Learn to see, see to learn*. Alexandria, VA: Assocation for the Supervision and Curriculum Development.

Burnett, R. (2004). *How images think*. Cambridge, MA: MIT Press.

Cabarga, L. (2001). *The designer's guide to global color combinations.* Cincinnati, OH: HOW Design Books.

Callahan, E. (2005). Cultural similarities and differences in the design of university Websites. *Journal of Computer-Mediated Communication,* 11(1), 239–273. Retrieved from http://jcmc.indiana.edu/vol11/issue1/callahan.html.

Carey, J. (1989). *Communication as culture: Essays on media and society.* New York: Routledge.

CBS News. (14 October 2011). Baby plays with iPad, is frustrated by paper mag. *The Early Show.* Retrieved from https://www.youtube.com/watch?v=uqF2gryy4Gs.

Charmaz, K. (2006). *Constructing Grounded Theory.* Thousand Oaks, CA: Sage.

Cooper, A. (2004). *The inmates are running the asylum: Why high tech products drive us crazy and how to restore the sanity.* Indianapolis: Sams Publishing.

Crane, D. (2002). Culture and globalization: Theoretical models and emerging trends. In D. Crane, N. Kawashima and K. Kawasaki (Eds.) *Global culture: Media, arts, policy, and globalization.* New York: Routledge. pp.1–25.

Csikszentmihalyi, M. (1991). *Flow: The psychology of optimal experience.* New York: Harper and Row.

Denmark. Retrieved from http://denmark.dk/en/meet-the-danes/.

Deuze, M. (2012). *Media life.* Cambridge, UK: Polity Press.

Donald, M. (1991). *The origin of the modern mind.* Cambridge, MA.: Harvard University Press.

Economist Intelligence Unit's quality-of-life index. (2005). *The Economist.* Retrieved from http://www.economist.com/media/pdf/QUALITY_OF_LIFE.pdf

Eisenstein, E. (2005). *The printing revolution in early modern Europe.* Cambridge: Cambridge University Press.

Ellul, J. (1964). *The technological society.* New York: Vintage Books.

Ess, C. and Sudweeks, F. (2005). Culture and computer-mediated communication: Toward new understandings. *Journal of Computer-Mediated Communication*, 11(1), 179–191. Retrieved from http://jcmc.indiana.edu/vol11/issue1/ess.html.

Faigley, L., George, D., Palchick, A. and Selfe, C. (2004). *Picturing texts.* New York: W. W. Norton & Company.

Finlay, V. (2003). *Color: A natural history of the palette.* New York: Random House.

Fuller, B. (1975). *Synergetics: Explorations in the geometry of thinking.* New York: Macmillan.

Geertz, C. (1973). *The interpretations of cultures.* New York: Basic Books.

Geertz, C. (1976). Art as a cultural system. *MLN 91*(6), 1473–1499.

Gell, A. (1992). *The anthropology of time: Cultural constructions of temporal maps and images.* Oxford; Berg.

Glaser, B. and Strauss, A. (1967). *The discovery of grounded theory: Strategies for qualitative research.* Chicago: Aldine Publishing Company.

Goffman, E. (1955). On face-work: An analysis of ritual elements of social interaction. *Psychiatry: Journal for the Study of Interpersonal Processes 18*(3), pp. 213–231. Reprinted in Goffman (2005, pp. 5–46)

Goffman, E. (1959). *The presentation of self in everyday life.* Garden City, NY: Doubleday.

Goffman, E. (1967). *Interaction ritual: essays in face-to-face behavior.* Random House. (2nd ed. with Joel Best, 2005). Aldine Transaction.

Goldwater, R. and Treves, M. (1945). *Artists on art.* New York: Pantheon Books.

Gorlick, A. (24 August 2009). Media multitaskers pay mental price, Stanford study shows. *Stanford Report.* Retrieved from http://news.stanford.edu/news/2009/august24/multitask-research-study-082409.html.

Graham, I. (2003). *A pattern language for web usability*. London: Addison-Wesley.

Greeno, J. G. (1994). Gibson's affordances. *Psychological Review (101)*2, pp. 336–342.

Guerrero, L.K., Devito J.A.,& Hecht M.L. (1999). *The Nonverbal Communication Reader: Classic and contemporary readings* (2nd ed.). Long Grove, IL: Waveland Press, Inc.

Hall, E. T. (1959). *The silent language*. New York: Doubleday/Premier Books.

Hall, E. T. (1969). *The hidden dimension*. New York: Anchor Books.

Hall, E. T. (1976). *Beyond culture*. New York: Anchor Books.

Hall, E. T. (1983). *The dance of life: The other dimension of time*. New York: Anchor Books.

Hall, E. T. and Hall, M. R. (1987). *Hidden differences: Doing business with the Japanese*. New York: Anchor Books.

Hall, S. (1997a). Old and new identities, old and new ethniticities. In A. D. King (ed.), *Culture, Globalization and the World-System* (pp. 41–68). Minneapolis: University of Minnesota Press.

Hall, S. (1997b). Introduction. In S. Hall (ed.) *Representation: Cultural Representations and Signifying Practices* (pp. 1–12). London: Sage.

Hardin, C. L. and Maffi, L. (1997). *Color categories in thought and language*. Cambridge: Cambridge University Press.

Hofstede, G. (1980). *Culture's consequences: International differences in work-related values*. Beverly Hills: Sage.

Hofstede, G. (1997). *Cultures and organizations: Software of the mind*. New York: McGraw-Hill.

Huffington Post. (18 March 2010; updated 25 May 2011). Google threatening to leave China over hacking, email leak. Retrieved from http://www.huffingtonpost.com/2010/01/12/google-threatening-to-lea_n_420857.html.

Huntington, S. (2012). The clash of civilizations? In F. Lechner and J. Boli (Eds.) *The globalization reader.* Malden, MA: John Wiley & Sons. pp. 37–44.

Innis, H. (1991). *The bias of communication.* Toronto: University of Toronto Press.

Kosslyn, S. (2006). *Graph design for the eye and mind.* New York: Oxford University Press.

Kosslyn, S. (2007). *Clear and to the point: 8 psychological principles for compelling Powerpoint presentations.* New York: Oxford University Press.

Kress, G. (2010). *Multimodality: A social semiotic approach to contemporary communication.* London: Routledge.

Kress, G. and van Leeuwen, T. (1996). *Reading images: The grammar of visual design.* London: Routledge.

Kocur, Z. (ed.) (2011). *Global visual cultures: An anthology.* Oxford: Wiley-Blackwell.

Koughan, F. and Rushkoff, D. (producers and writers). (February 18, 2014). Generation Like. In D. Fanning (executive producer) Frontline. New York: Left/Right Docs.

Larsson, T. (2001). *The race to the top: The real story of globalization.* U.S.: Cato Institute.

Levin, R. (1998). *A geography of time: The temporal misadventures of a social psychologist, or how every culture keeps time just a little bit differently.* New York: Basic Books.

Lewis, R. (1999). *When cultures collide: Managing successfully across cultures.* London: Nicholas Brealey Publishing.

Lievrouw, L. A. and Livingstone, S., Eds. (2006). *The handbook of new media.* London: Sage Publications.

Lipton, R. (2002). *Designing across cultures.* Cincinnati, OH: HOW Design Books.

Logan, R. (2007). *The extended mind: The emergence of language, the human mind, and culture.* Toronto: University of Toronto Press.

Logan, R. (2010). *Understanding new media: Extending Marshall McLuhan.* New York, Peter Lang.

Lule, J. (2012). *Globalization and media: Global village of Babel.* Lanham, MD: Rowman & Littlefield.

Machin, D. and van Leeuwen, L. (2007). *Global media discourse.* New York: Routledge.

Manovich, L. (2001). *The language of new media.* Cambridge, MA: MIT Press.

McCandless, D. (2009). *Information is beautiful.* London: Collins.

McCreight, T. (1996). *Design language.* Portland, ME: Brynmorgen Press.

McLuhan, M. (1964, 1994, 2002). *Understanding media.* Cambridge: MIT Press.

McLuhan, M. and Powers, B. (1989) *The global village: Transformations in world life and media in the 21st century.* New York: Oxford University Press.

Menzies, H. (2005). *No time: Stress and crisis of modern life.* Vancouver: Douglas & McIntyre.

Mirzoeff, N., Ed. (1998). *The visual culture reader.* London: Routledge.

Moggridge, B. (2007). *Designing interactions.* Cambridge, MA: The MIT Press.

Mullet, K. and Sano, D. (1995). *Designing visual interfaces: Communication oriented techniques.* Mountain View, CA: Prentice Hall.

Mumford, L. (1934, 1962). *Technics and civilization.* San Diego: Harcourt Brace & Company.

Nederveen Pieterse, J. (2009). *Globalization and culture.* New York: Rowman & Littlefield Publishers, Inc.

Newton, R. (2004). *Galileo's pendulum: From the rhythm of time to the making of matter.* Cambridge, MA: Harvard University Press.

Nielsen, J. (2006). F-Shaped pattern for reading web content. *Jakob Nielsen's Alertbox*. Nielsen Norman Group. Retrieved from http://www.nngroup.com/articles/f-shaped-pattern-reading-web-content/

Nielsen, J. (2012, January 4). Usability 101: Introduction to Usability [Online]. *Nielsen Norman Group*. Retrieved from http://www.nngroup.com/articles/usability-101-introduction-to-usability/.

O'Connell, R. and Mateus, K. (September 2004). Regional newspaper website design. *Mequoda Group*. Retrieved from http://mequoda.com/wp-content/uploads/pdfs/MequodaLibraryReportIndex.pdf

O'Connell, R. (2010). Website usability. In S. Josephson, S., S. B. Barnes and M. Lipton (Eds), *Visualizing the Web* (pp. 123–140). New York: Peter Lang.

Ophir, E., Nass, C. and Wagner, A. D. (2009). Cognitive control in media multitaskers. *Proceedings of the National Academy of Sciences*. Retrieved from http://www.pnas.org/content/early/2009/08/21/0903620106.full.pdf+html

Outing, S. (2004, 2011). Eyetrack III: What news websites look like through reader's eyes. *Poynter*. Retrieved from http://www.poynter.org/uncategorized/24963/eyetrack-iii-what-news-websites-look-like-through-readers-eyes/

Peterson, L. K. (2000). *Global graphics: Color*. Gloucester, MA: Rockport Publishers.

Pettigrew, T. F. (1958). The measurement and correlates of category width as a cognitive variable. *Journal of Personality* (26) 532–544.

Prosser, J., Ed. (1998). *Image-based research*. London: Routledge.

Rhyne, R. F. (2003). *The act of choosing: A context-matching theory, and its practical implications*. Lincoln, NE: iUniverse.

Robertson, R. (2012). Globalization as problem. In F. Lechner and J. Boli (Eds.) *The globalization reader*. Malden, MA: John Wiley & Sons. pp. 88–94.

Rokeach, M. (1968). *Beliefs, attitudes, and values.* San Francisco: Jossey-Bass.

Rokeach, M. (1973). *The nature of human values.* New York: The Free Press.

Rose, G. (2001, 2007, 2012). *Visual methodologies: An introduction to the interpretation of visual materials.* London: Sage Publications.

Rushkoff, D. (2013). CNN: NSA phone snooping, a new kind of creepy. *Rushkoff.* Retrieved from http://www.rushkoff.com/blog/2013/6/6/cnn-nsa-phone-snooping-a-new-kind-of-creepy.html

Samovar, L., Porter, R. and Stefani, L. (1998). *Communication between cultures.* 3rd Ed. Belmont, CA: Wadsworth Publishing Company.

Saussure, F. de ([1916] 1983): *Course in general linguistics* (trans. Roy Harris). London: Duckworth

Schriver, K. A. (1997). *Dynamics in document design: Creating text for readers.* New York: John Wiley & Sons.

Segev, E., Ahituv, N., and Barzilai-Nahon, K. (2007). Mapping diversities and tracing trends of cultural homogeneity/heterogeneity in cyberspace. *Journal of Computer-Mediated Communication*, 12(4), 1269–1297. Retrieved from http://jcmc.indiana.edu/vol12/issue4/segev.html.

Singh, N. and Pereira, A. (2005). *The culturally customized Website.* Oxford: Elsevier.

Smith, K., Moriarty, S., Barbatsis, G. and Kenney, K. (2005). *The handbook of visual communication.* Matwah, NJ: Lawrence Erlbaum Associates, Publishers.

Spool, J., Perfetti, C. and Brittan, D. (2004). Designing for the scent of information. *UIE Fundamentals Reports.* Retrieved from http://www.uis.edu/webservices/wp-content/uploads/sites/8/2013/02/Designing_for_Scent.pdf

Steiner, P. (1993, July 5). *On the Internet, nobody knows your a dog* [cartoon]. *The New Yorker.*

Sturken, M. and Cartwright, L. (2009). *Practices of looking: An introduction to visual culture.* New York: Oxford University Press.

Thorlacius, L. (2004). Aesthetics and function in Web design. *CADE2004 Web Proceedings of Computers in Art and Design Education Conference*, M. Agger Eriksen, L. Malmborg, J. Nielsen (eds.) Copenhagen Business School, Denmark and Malmö University, Sweden, 29 June–1 July 2004.

Thorlacius, L. (2007). The role of aesthetics in Web design. *Nordicom Review* 28 (1), 63–76.

Tidwell, J. (2011). *Designing interfaces.* Sebastopol, CA: O'Reilly Media.

Todorović, D. (2007). W. Metzger: Laws of Seeing. *Gestalt Theory, 28,* 176–180.

Tomlinson, J. (2012). Cultural imperialism. In F. Lechner and J. Boli (Eds.) *The globalization reader.* Malden, MA: John Wiley & Sons. pp. 343–351.

United Nations, Department of Economic and Social Affairs, Statistics Division, Demographic and Social Statistics Branch. (2004). United Nations *Demographic Yearbook* review: National reporting of international migration data; implications for international recommendations. Retrieved from: http://unstats.un.org/unsd/demographic/products/dyb/techreport/migration.pdf.

Van Leeuwen,, T. (2005). *Introducing social semiotics.* London: Routledge.

Victor, D. (2013). Doing business in Argentina. *Encyclopedia of Business, 2nd ed.* Retrieved from http://www.referenceforbusiness.com/encyclopedia/A-Ar/Argentina-Doing-Business-in.html

Ware, C. (2008). *Visual thinking for design.* Burlington, MA: Morgan Kaufmann Publishers.

What is Scandinavian design? Fashionising.com. Retrieved from http://www.fashionising.com/trends/b--future-of-scandinavian-design-12713.html/2

What is wikileaks? Wikileaks.org. Retrieved from http://wikileaks.org/About.html

Williams, R. and Tollett, J. (2005). *The non-desginer's web book, 3rd ed.* Berkeley, CA: Peachpit Press.

Wuertz, E. (2005). A cross-cultural analysis of websites from high-context and low-context cultures. *Journal of Computer-Mediated Communication*, 11(1), article 13. Retrieved from http://jcmc.indiana.edu/vol11/issue1/wuertz.html.

Index

D

F

G

L

M

O

P

Q

R

X

Y

Z

Visual COMMUNICATION

Susan B. Barnes, *General Editor*

Visual communication is the process through which individuals in relationships, organizations, and cultures interpret and create visual messages in response to their environment, one another, and social structures. This series seeks to enhance our understanding of visual communication, and explores the role of visual communication in culture. Topics of interest include visual perception and cognition; signs and symbols; typography and image; research on graphic design; and the use of visual imagery in education. On a cultural level, research on visual media analysis and critical methods that examine the larger cultural messages imbedded in visual images is welcome. By providing a variety of approaches to the analysis of visual media and messages, this book series is designed to explore issues relating to visual literacy, visual communication, visual rhetoric, visual culture, and any unique method for examining visual communication.

For additional information about this series or for the submission of manuscripts, please contact Dr. Barnes at *susanbbarnes@gmail.com.*

To order other books in this series, please contact our Customer Service Department:

(800) 770-LANG (within the U.S.)
(212) 647-7706 (outside the U.S.)
(212) 647-7707 FAX

Or browse online by series at www.peterlang.com.